POINT CLOUD DATA FUSION FOR ENHANCING 2D URBAN FLOOD MODELLING

cover photograph by Roisri (2011)

POINT CLOUD DATA FUSION FOR ENHANCING 2D URBAN FLOOD MODELLING

DISSERTATION

Submitted in fulfilment of the requirements of
the Board for Doctorates of Delft University of Technology
and
of the Academic Board of the UNESCO-IHE
Institute for Water Education
for
the Degree of DOCTOR
to be defended in public on
Wednesday June 14, 2017 at 10:00 hours
in Delft, the Netherlands

by

Vorawit MEESUK

Master of Science in Remote Sensing and Geographic Information System,
Khon Kaen University, Thailand
Bachelor of Science in Computer Science and Technology,
Rajamangala Institute of Technology, Thailand

born in Bangkok, Thailand

This dissertation has been approved by the
promotor: Prof. dr. ir A.E. Mynett and
copromotor: Dr. Z. Vojinović

Composition of the doctoral committee:

Chairman	Rector Magnificus Delft University of Technology
Vice-Chairman	Rector UNESCO-IHE
Prof. dr. ir A.E. Mynett	UNESCO-IHE / Delft University of Technology, promotor
Dr. Z. Vojinović	UNESCO-IHE, copromotor

Independent members:

Prof. dr. ir N.J. van de Giesen	Delft University of Technology
Prof. dr. R. Ranasinghe	UNESCO-IHE / University of Twente
Prof. dr. D. Savic	University of Exeter, United Kingdom
Dr. S. Weesakul	Asian Institute of Technology, Thailand
Prof. dr. D.P. Solomatine	Delft University of Technology / UNESCO-IHE, reserve member

This research was conducted under the auspices of the Graduate School for Socio-Economic and Natural Sciences of the Environment (SENSE).

CRC Press/Balkema is an imprint of the Taylor & Francis Group, an informa business

Published by:
CRC Press/Balkema
PO Box 11320, 2301 EH Leiden, the Netherlands
e-mail: Pub.NL@taylorandfrancis.com
www.crcpress.com – www.taylorandfrancis.com
ISBN 978-1-138-30617-2 (Taylor & Francis Group)

to my dear family

Summary

Modelling urban flood dynamics requires detailed knowledge of a number of complex processes. Numerical simulation of flood propagation requires proper understanding of all processes involved. Also, special care has to be taken to adequately represent the urban topography, taking into account typical urban features like underpasses and hidden alleyways. Incorrect input data will affect the numerical model results, which may lead to inadequate flood-protection measures or even catastrophic flooding situations. This thesis explores how to include particular urban topographic features in urban flood modelling.

Aerial Light Detection and Ranging (LiDAR) systems offer opportunities for achieving good quality topographic data, with less fieldwork on the ground. Even though aerial LiDAR data have long been used in many applications, conventional top-view LiDAR data can not quite capture some hidden urban features. However, recent improvements in Structure from Motion (SfM) techniques provide opportunities to achieve improved quality topographic data by using multiple viewpoints, including side-view data.

This dissertation explores insights into the capabilities of assimilating SfM point cloud data by using multi-source views as input for enhancing 2D urban flood modelling. Side-view SfM point cloud data are collected and merged with conventional top-view LiDAR point cloud data to create novel Multi-Source Views (MSV) topographic data. The main objectives of this research are (i) to provide insight into the capabilities of using computer-based environments; (ii) to explore the benefits of using the new MSV data; (iii) to enhance 2D model schematizations; (iv) to compare simulated results using new MSV data and conventional top-view LiDAR data as input for urban flood models; and (v) to help developing flood-protection measures.

Three conventional top-view LiDAR DEMs, viz. (i) LiDAR Digital Surface Model (DSM); (ii) LiDAR Digital Terrain Model (DTM); (iii) LiDAR Digital Building Model (DBM+); and (iv) the new MSV Digital Elevation Model (DEM) and used as input to set up 2D model schematizations for two case studies: the 2003 flood event in Kuala Lumpur, Malaysia; and the 2011 flood event in Ayutthaya, Thailand. It was observed that by merging side-view SfM data with conventional top-view LiDAR data into the new MSV dataset, most key components of urban features could be incorporated. When applying MSV-DEM as input, some specific components of complex urban features (e.g. retention walls under trees, pathways under arches, and kerbs under sky-train tracks) could be taken into account for enhancing 2D model schematics. Making use of the new MSV topographic input data led to promising results by enhancing 2D model schematizations. Simulation results appear to represent more realistic flood dynamics, especially in complex cities.

Findings showed that simulation results using conventional top-view LiDAR-DSM as input show the least flood inundation areas and results contained many dry areas. This is because conventional LiDAR-DSM does not capture hidden urban features like underpasses, high trees, overarching structures, which behave as obstacles in 2D model schematics. When applying extended top-view LiDAR-DBM+ and the newly developed MSV-DEM as input, simulation results showed more inundation areas closer to reality.

In the case study for Kuala Lumpur, the enhanced simulation results revealed missing inundation areas since floodwaters can now freely flow through sky train tracks. In the Ayutthaya case, merging different views of point cloud data showed that SfM data provide opportunities for better analysing historical flood magnitudes and their elevation benchmarks. The extracted elevation data showed good agreement with conventional data observed by land surveys, which looks promising for verifying numerical simulation results. For urban-flood mapping, findings showed that when using LiDAR-DTM as input, the flood maps for floodwater depth and flood inundation area can help developing local and regional flood-protection measures.

This research shows that high-performance computational simulation environments can play a significant role in shortening simulation times for high-resolution urban-flood modelling. On a stand-alone computer, simulation models can perform adequately for simple cases. However, real powerful parallel computing can far better show increased performances of advanced models when using cloud-computing services, like from SURFsara in Amsterdam, as used in this thesis.

Samenvatting

Het modelleren van overstromingen in stedelijke gebieden vereist gedetailleerde kennis van verschillende complexe processen, waaronder numerieke simulatie van overstromingen en het juist weergeven van stedelijke topografie met zijn specifieke eigenschappen van onderdoorgangen rond gebouwen en verscholen steegjes. Als deze niet goed worden meegenomen in de schematisaties, dan zullen de numerieke modeluitkomsten geen juiste weergave bieden, wat weer kan leiden tot onjuiste maatregelen om overstromingen te voorkomen, of zelfs tot catastrofale situaties. Dit proefschrift richt zich op de vraag hoe de topografie in stedelijke gebieden het best kan worden meegenomen bij numerieke modelvorming.

LiDAR (Aerial Light Detection and Ranging) systemen bieden de mogelijkheid om een goede kwaliteit topografische kaart samen te stellen zonder ingewikkelde en tijdrovende metingen op de grond. Hoewel LiDAR al geruime tijd wordt toegepast blijkt dat bovenaanzichten niet goed in staat zijn om bepaalde 'verborgen' doorgangen mee te nemen. Met nieuwe ontwikkelingen op het gebied van Structure from Motion (SfM) technieken is het mogelijk om topografische kaarten te verbeteren door gebruik te maken van zijaanzichten.

In dit proefschrift wordt nagegaan hoe SfM waarnemingen kunnen worden gebruikt om standaard 2D overstromingsmodellen voor stedelijke gebieden te verbeteren. Daartoe zijn zijaanzichten gebruikt in combinatie met bovenaanzichten om een nieuwe Multi-Source Views (MSV) topografische kaart te construeren. De belangrijkste doelstellingen van dit onderzoek zijn om (i) inzicht te verkrijgen in de mogelijkheden die computer-ondersteunde omgevingen bieden; (ii) de voordelen na te gaan van het gebruik van MSV gegevens; (iii) daarmee 2D modelschematisaties te verbeteren; (iv) om de stromingsberekeningen van deze niewe MSV aanpak te vergelijken met overstromingssimulaties die alleen zijn gebaseerd op traditionele bovenaanzicht LiDAR topografie; (v) om op basis hiervan veiligheidsmaatregelen voor te stellen tegen overstromingen.

Drie traditionele manieren van bovenaanzichten, (i) LiDAR Digital Surface Model (DSM); (ii) LiDAR Digital Terrain Model (DTM); (iii) LiDAR Digital Building

Model (DBM+); worden vergeleken met (iv) de nieuwe MSV Digital Elevation Model (DEM) aanpak om stedelijke gebieden te schematiseren voor twee toepassingen: (i) de overstromingen in Kuala Lumpur, Maleisië in 2003; en (ii) de overstromingen in Ayutthaya, Thailand in 2011. Door gebruik te maken van zijaanzichten in combinatie met traditionele bovenaanzichten bleek dat de nieuwe MSV dataset de meeste stedelijke topografische bijzonderheden goed kon weergeven. Wanneer MSV-DEM als invoer wordt gebruikt kunnen bijzonderheden als muren verscholen onder bomen, onderdoorgangen verstopt onder gebouwen, stoepranden en verhogingen niet zichtbaar onder onder sky-train, etc. goed worden meegenomen in de 2D overstromingsberekeningen. Deze nieuwe aanpak lijkt dan ook veelbelovend. De rekenresultaten sluiten nauwer aan bij de waarnemingen, in het bijzonder bij complexe stedelijke configuraties.

Schematisaties gebaseerd op alleen LiDAR-DSM bovenaanzichten leidden tot de minste overstromingen en relatief veel droge gebieden. Dit komt omdat in deze gevallen geen verborgen doorgangen in de topografie kunnen worden meegenomen waardoor het water (ogenschijnlijk) wordt geblokkeerd. Als de specifieke eigenschappen van gebouwen in de LiDAR-DBM+ wordt meegenomen, dan komen de resultaten beter overeen met waarnemingen.

Bij toepassing in Kuala Lumpur kon het overstromingsmodel aanzienlijk worden verbeterd bij met name de bovengrondse spoorlijn die bij een naieve interpretatie als 'spoordijk' werd gezien. In het geval van Ayutthaya kon de SfM techniek voor zijaanzichten worden gebruikt om een historisch overzicht te construeren van de waterdieptes tijdens de grote overstromingen van 2011. De resultaten vertoonden goede overeenkomst met de cartografische waarnemingen die de Thaise overheid had laten uitvoeren. Op basis hiervan kon ook geadviseerd worden over te nemen lokale en regionale maatregelen om overstromingen in de toekomst tegen te gaan.

Het onderzoek in dit proefschrift laat zien dat geavanceerde computer-omgevingen een belangrijke rol kunnen spelen bij het versnellen en interpreteren van gedetailleerde numerieke modelsimulaties in complexe stedelijke gebieden. Voor eenvoudige toepassingen volstaat vaak een enkele computer. Echter, gedetailleerde berekeningen kunnen pas worden verkregen door gebruik te maken van krachtige parallelle rekentechnieken en cloud-computing services, zoals die van SURF-sara in Amsterdam, die voor dit proefschrift zijn gebruikt.

Acknowledgements

Time has come to remind me that my long PhD journey had carried out through six years at UNESCO-IHE in Delft. Without doubt, this research would not have been possible without help and support from the kind people around me, whom of which are possible to give particular mention hereafter.

First, I thank the sponsor: Royal Thai Government Scholarship for financial support; Hydro and Agro Informatics Institute (HAII), Ministry of Science and Technology (MOST), Thailand for their support through the years. I am very grateful to Dr. Royol Chitradon for trusting and giving me opportunities to grow at HAII/MOST.

My sincere gratitude goes to Prof. Arthur Mynett, my promotor, for his continuous support. Finalising this research even during Christmas and New Year holidays, manifested his commitment as a supervisor to my research. Grateful thanks also go to Dr. Zoran Vojinović, my copromotor, for his valuable input and software support to this research. He was the first to introduce me to PhD experience at one of the best institutes for water education, UNESCO-IHE in Delft, the Netherlands. My sincere gratitude goes to my promotor and copromotor for their continued support throughout my entire PhD journey.

For the case study fieldwork of Kuala Lumpur, Malaysia, I would like to thank the Department of Irrigation and Drainage (DID), Malaysia for providing valuable data. The kind support from Dr. Abdullah and Ms. Aziz for their advice and assistance with data collection and data processing is greatly appreciated.

For the case study fieldwork of Ayutthaya, Thailand, I would like to thank the Geo-Informatics and Space Technology Development Agency (GISTDA/MOST), Thailand, for providing valuable data. Thanks go to Mr. Nakmuenwai and Mr.

Sansena for providing assistance in data processing, as well as to Mr. Thammsittirong and the AIT team for their helpful assistance in modelling setup. The kind support received from the Department of Science Service (DSS/MOST), Thailand, including Dr. Kulvanit, Mr. Praeknokkeaw, Mr. Phosuk, Mr. Surattisak in modifying mobile units for street-view surveys is greatly acknowledged. Many thanks go to Mr. Wangkiat from the Institute of Field Robotics (FiBO) and Mr. Kanyawararak, Mr. Sa-Ngiam, Mr. Jindasee, Mr. Chantee, and Mr. Sae-tear from HAII, for supporting my field work.

The research work presented here has received partial funding from the European Union Seventh Framework Programme (FP7/2007-2013) under Grant agreement n° 603663 for the research project PEARL (Preparing for Extreme And Rare events in coastaL regions).

This research would not have been possible without MIKE by DHI™ for providing their numerical models. Many thanks also go to Mr. Astudillo, Ms. Danezi, and their team for the excellent HPC cloud computing facilities of SURFsara web services in Amsterdam, the Netherlands.

Many thanks go to all my friends in the Netherlands that made me feel at home. I would like to extend my gratitude to all of them and their families. Many thanks in particular go to the people at UNESCO-IHE including Dr. Corzo, Dr. Paron, Mr. Ceton, Mr. van Nievelt, and Mr. Kleijn, for helping me with software recommendations and their useful IT assistance. Thanks to Ms. Tonneke Morgenstond, for providing me a workplace with unforgettable view of the Oude Delft. Many thanks to Ms. Jolanda Boots, Ms. Sylvia van Opdorp, Ms. Anique Karsten, and Ms. Mariëlle van Erven for all their kind assistance and useful recommendations for making life in the Netherlands feel like home.

Table of content

CHAPTER 1
Introduction

Predicting urban flood dynamics is extremely complex. Incorrectly analysing data and model results may lead to inadequate flood-protection measures or even lead to more catastrophic situations. Properly analysing topographic data and key components of urban features is crucial for complex flood analyses. Nowadays remote sensing technologies bring new opportunities to achieve high quality topographic data. Amongst these technologies, aerial light detection and ranging (LiDAR) system can adequately be used to obtain topographic data for peri-urban and complex urban areas in a matter of weeks or days. These conventional LiDAR topographic data are obtained from top views and have long been used in many applications. However, in more complex cities, some urban features are difficult to trace and these features are often omitted in conventional top-view LiDAR data. More qualitative analyses of topographic data obtained from different sources and different viewpoints should result in better urban feature representations, flood model schematizations, and flood simulation results. This dissertation describes a way to make use of Structure from Motion (SfM) techniques to obtain point cloud data to create different-source views of topographic data for enhancing urban-flood models. In this chapter, principles of urban flooding and topographic input data are introduced in Section 1.1 and 1.2, resp. Overall objectives and research questions are given in Section 1.3. The outline of this dissertation is also shown in Section 1.4.

1.1 Urban flooding

Through history, human settlements are located nearby rivers and coasts. Fresh and seawater environments ensure accesses of food and water for human daily consumption. In modern world, living close to water resources also brings more opportunities of transporting, trading, agriculture, etc. to people. Many cities are centres of economics and politics, magnetising inflows of people and investments. Big cities commonly have more populations that may lead to scarcities of living spaces. Such big cities often expand its centre to low-lying areas (De Sherbinin et al., 2007) and sometimes these areas are below sea level. Almost all these circumstances could be counted that such city's areas are prone to flooding.

Experiences during last decades revealed that most of modern cities were affected by natural disasters, of which over 70% were flood-related incidents (EM-DAT, 2014). On the basis of GDP per capita and population densities, a projected exposure of flooding in cities will be extensively increasing by 2050 with total losses more than USD 158 trillion (Jongman et al., 2012). According to a research of Balmforth et al. (2006) that urban floods were grouped into four broad categories: (i) fluvial or river floods, (ii) pluvial or local floods, (iii) tidal or coastal floods, and (iv) groundwater floods. Often floods at any time and location can also arise from these different category combinations. When residential, commercial, and industrial developments were settled in such flood-prone areas, they are obviously vulnerable to flood disasters (Mynett & Vojinović, 2009; Singh et al., 2016).

Although losses cannot be avoided when a major flood occurs, preparedness could considerably minimise the number of lives lost and reduce damage costs from flooding. Minimising flood damage costs need balanced approaches in structural and non-structural measures that should be considered with intensive care (Green, 2004). Lacking or overlooking information support can lead to impractical measures and inappropriate decisions. For developing cities, decent quality of information support is commonly a lack of access, outdated observations, or even never been surveyed. Such adequate information support (e.g. topographic and bathymetry data, hydrological data, and simulation results) is vital for creating new flood-protection measures also for visualising effects of proposed measures during public hearing processes and before implementing those new measures into reality.

In complex cities, behaviour of floodwater flows and their routeing processes is crucially associated with existing topographic conditions. Changes in topographic conditions may have a profound impact on overland flow processes (Mitasova et al., 2011). However, these flow directions are not only controlled by topographic terrains (e.g. elevations and slopes) but also conveyed by complex urban structures (e.g. pathways and obstacles; Fig. 1-1).

Fig. 1-1.   Pathum Thani City residents evacuated through flooded streets on October 22nd, 2011 in Thailand (source by Berehulak, 2011)

On this point, numerical urban-flood models as hydroinformatics supporting tools can play a significant role in replicating flood dynamics and evaluating effects of proposed measures for different scenarios. In reality, even though some flood may have insignificant changes after a very long time during the flood event, their initial stage of the flood event could have significant dynamic flows, which rapidly change during a short time. Scale factors of developing appropriate qualities for 2D model schematics are crucially depending on time scales and space scales in order to simulate flood either dynamic or even static parts. Therefore these simulation results should properly represent evolutions of flood extents, floodwater depths, rates of rising flood levels, and lead time prior to human activities being interrupted by flooding (Vojinović & Abbott, 2012). This means that the choices of model schematic setups have to be addressed in great detail for each flood situation.

By making use of numerical models, simplifying some complexities of cities can be sufficient for simple flood simulations. Simplification like one-dimensional (1D) numerical models are simple and safe for decision-making and also have been practically used for a long time already. The 1D models have been typically used because they are relatively easy for modelling setups, calibrations, and explanations. For some situations, the 1D models may be impractical when floodwaters exceed capacities of confined conduits. Extending to quasi 2D approaches could be feasible for handling such situations. Using 1D models remains valid particularly in simulating flood flow dynamics for a single direction.

Even though simulating flow dynamics in more than one direction can be calculated in both 2D and 3D models, simulating these flows in 2D models seem to be more practical and straightforward for most cases of urban flood predictions. In 2D models, an assumption of nearly horizontal flows is indicated in the shallow-water equations, which allow considerable simplifications in mathematical formulations and numerical solutions. This assumption not only considerably simplifies analyses but also yields reasonable explanations and representations. However, vertical flows in 2D models are commonly omitted, but still considerably concerned in 3D models.

Nowadays, coupled 1D-2D models are commonly used for simulating flood dynamics of conjunctive areas between confined conduits using 1D models and urban floodplains using 2D models. Calibration of these physically-based models is recursive adjustment processes which often uses parameter values to achieve the best fit of simulation results, close to observed measurements as much as possible. Calibrations should be handled with care to avoid unreasonably or over adjusting parameters falling into the trap of force-fitted model. The values of each parameter could be achieved by analysing and experimentally testing in laboratories or even quantitatively defining as descriptions. Some guidelines may be given in a range of appropriate values, but precise values can vary from case to case. Exceeding these limitations may lead to incorrectly simulated results. Verifying these models are needed for evaluating such simulated results continue to be reasonable for another set of observed measurements.

Time-series data of water levels and discharges are commonly observed and used for verification and calibration processes. Systematic gauges are typically kept records and often installed only for some main rivers (Gee et al., 1990; Bates et al., 1992). It may be more difficult to find these gauges distributedly installed for whole river networks or even kept such records overland flows for floodplains. A lack of these observed measurements is still the main issue for some cities. Some researchers have further explored capabilities of post-flood analyses as alternative approaches for flood observations. Many research showed that flood extent maps were created by using aerial photos (Connell et al., 2001; Overton, 2005; Yu & Lane, 2006), airborne data, and satellite images (Horritt, 2000; Brivio et al., 2002; Rosenqvist et al., 2002; Townsend & Foster, 2002; Bates et al., 2006). These flood extent maps can adequately represent boundaries of inundation areas. However, they are often deficient in representing evolutions of flood propagations.

When applying post-flood analyses, the analysed flood peaks could be applied as benchmark and incorporated into model calibration, verification, and validation processes (Han et al., 1998; Dutta et al., 2000; Hervouet, 2000; Hsu et al., 2000a; Hesselink et al., 2003; Romanowicz & Beven, 2003). Typically, peak elevations of flood watermarks are measured by conventional land surveying (Aronica et al., 2002; Werner et al., 2005). However, measuring such watermarks using conventional land surveys still consumes a lot of labour intensive efforts. Owing to this, some methods for getting watermark information (e.g. Gaume & Borga, 2008) are explored here using SfM technique. Making use of post-flood analyses from watermarks could also enhance quality of urban flood model results and improve better understanding insight into processes of urban flood dynamics, especially for complex cities.

1.2 Topographic input data for urban flood modelling

Conventional land surveys have long been used to obtain decent accuracy of topographic data. However, these conventional land surveys use intensive works, yet it cannot provide high details of topographic data. Even though high-resolution topographic data are becoming more require these days, achieving such data at urban feature resolutions is still difficult (Vojinović & Abbott, 2012).

Ever since remote sensing technologies have emerged, topographic data can be obtained without contact. During the last decades, remote sensing technologies have dramatically improved quality of topographic data. Amongst these technologies, aerial light detection and ranging (LiDAR) systems can provide high accuracies of topographic data in sub metre resolutions. The aerial LiDAR surveys have been commonly used to obtain top-view topographic data for a long time already in ranges of applications.

For urban flood management applications, many researchers (Marks & Bates, 2000; Horritt & Bates, 2001; Haile & Rientjes, 2005b; Wright, 2005; Fewtrell et al., 2011; Sampson et al., 2012) showed that improvements in model resolutions using top-view LiDAR data can have considerable effects on inundation extents and timing prediction results. Availablilities to these top-view LiDAR data can still be found only in a few cities, yet achieving such data often spends many investment costs. Moreover, these top-view LiDAR data have difficulties in representing some key components of urban features, i.e. vertical structures and low-level structures. These urban features are typically hidden underneath overreaching structures or trees, which are easily neglected in both conventional land surveys and conventional aerial LiDAR surveys. However, these missing key components can play an important role in flood dynamics especially in complex cities. Some of these urban features may allow floodwaters flowing through, while other features may obstruct or divert floodwater flows differently.

Nowadays, huge improvements in topographic data acquisitions bring several alternatives for achieving proper quality topographic data from different remote sensing sources. Structure from Motion (SfM) techniques, for example, could be counted as another outstanding tool, which can provide more accessibilities for achieving topographic data in high resolutions and high accuracies, with promising investment costs. In SfM surveys, normal compact digital cameras as a surveying tool can be simply mounted on several surveying platforms for obtaining topographic data from different viewpoints (e.g. side and top viewpoints). Notably, recent studies introduced benefit of using different sources of topographic data as input for enhancing quality of flood simulation results (Fewtrell et al., 2011; Sampson et al., 2012). Owing to this, qualitative analysis in different-sources views of topographic data could bring better quality of urban flood schematizations and could reveal more benefits to their simulation results.

Advances in 2D urban-flood models nowadays offer potentials to predict local flood patterns (e.g. flood inundation areas and flood depths) closer to reality, also simulate flood flow dynamics (e.g. flood velocities and flood flow routeings) more accurate with much better efficient computational cost. Developing such appropriate qualities for 2D model schematics are crucially depending on accuracies and computational efficiencies. However, a way to get better understanding of developing, applying, and making use of merging point cloud data from different-source views for enhancing urban-flood models are still challenging.

1.3   Objectives and research questions

Overall objectives of this research are to explore state of the art in modelling key components of urban features, which are crucial for urban flood analyses. Such key components could play a significant role for complex flood dynamics and inundation behaviours in cities. Decent quality topographic data could represent more qualitative details of urban features. Even though aerial surveys have long been used to obtain LiDAR data, these conventional top-view LiDAR data still have their difficulties to represent such key components.

The more qualitative analysis of topographic data from different sources and different viewpoints could bring better indications of key components and could improve flood model schematizations also their simulation results. In this research, a way to get better understanding of analysis for different-source views topographic data and a way to make use of enhancing such data as input for urban flood models are further explored.

Huge improvements in SfM technology bring new opportunities to achieve improved quality topographic data obtained from different viewpoints. In this research, the SfM technologies are applied for side-view surveys. Overlapping photos taken from side-view surveys are used to create side-view SfM topographic data, which are reconstructed by using open software (e.g. VisualSFM and PMVS-CMVS), web-based services (e.g. PhotoSynth), or commercial software (e.g. PhotoScan). Furthermore, a new concept of merging side-view SfM data with conventional top-view LiDAR data is introduced for creating a novel multi-source views (MSV) topographic data. This new MSV data could be appropriately used as input for complex flood models. In order to replicate complex flood situations like urban floods, utilising 1D numerical models may not be valid for some situations (e.g. lateral flows). Adopting 2D models could be more appropriate for simulating such complex floods and seems to be more straightforward than applying sophisticated 3D models.

In hypothetical cases, we explore capabilities of using new MSV topographic data as input for 2D models. We then evaluate their simulation results of complex flood situations by comparing simulation results using new MSV data and conventional top-view LiDAR data as input. We further explored which topographic data should be capable of enhancing 2D model schematizations and replicating more realistic flood dynamics, especially for complex cities. Example implementations are carried out for two case studies: (i) the 2003 flood event in Kuala Lumpur, Malaysia and (ii) the 2011 flood event in Ayutthaya, Thailand. In this respect, advances in coupled 1D-2D models using MIKE FLOOD™ commercial software are promising to replicate such complex flood situations.

This dissertation aims to provide insights into the capabilities of using advances in computer-based environments of urban flood models for enhancing model schematizations and exploring effects of proposed flood-protection measures. Owing to this, the five specific research questions are formulated as follows:

1. *What are the limitations of using conventional top-view LiDAR data as input for urban flood modelling?*
2. *What are the benefits of using SfM technique for creating topographic data?*
3. *How to fuse 3D point cloud data for constructing proper elevation maps?*
4. *How can 3D point cloud data be used for enhancing 2D urban flood models?*
5. *How can computer-based environments help developing flood-protection measures?*

1.4 Dissertation outline

Short descriptions as guidance notes for each chapter are given as follows:

Chapter 1 outlines concept ideas, overviews general objectives, and formulates main research questions related to the dissertation.

Chapter 2 reviews fundamental backgrounds of urban flood modelling. Hydrodynamic principles in 1D, 2D, and coupled 1D-2D numerical models are given. Issues concerning complex urban flood modelling are discussed.

Chapter 3 reviews basic backgrounds of conventional top-view LiDAR surveys. The evolution in topographic elevation data acquisition is described. Top-view LiDAR data acquisition and top-view LiDAR data simplifications are given. Issues concerning conventional top-view LiDAR data are discussed.

Chapter 4 introduces side-view SfM surveys. The key components of urban features, which can be detected and extracted from side-view SfM topographic data are explored. Issues concerning side-view SfM data are discussed.

Chapter 5 develops the novel MSV topographic data by merging side-view SfM data with conventional top-view LiDAR data. Evaluating the use of new MSV topographic data and conventional top-view LiDAR data as input for enhancing urban flood models are discussed.

Chapter 6 implements new MSV topographic data as input for simulating a real flood event. The 2003 urban flood event in Kuala Lumpur, Malaysia is used for the first case study. The simulated results using new MSV data and conventional top-view data are compared, evaluated, and discussed.

Chapter 7 explores the use of flood watermarks extracted from side-view SfM data. These extracted watermarks are used as benchmark for verifying simulated floodwater-depth results. The 2011 urban flood event in Ayutthaya, Thailand is used for the second case study. The simulated results using new MSV data and conventional top-view data are compared, evaluated, and discussed.

Chapter 8 recommends further flood-protection measures for the Ayutthaya Island case study. Several scenarios using urban flood simulation as hydro-informatics supporting tools are evaluated and discussed.

Chapter 9 further outlooks of multi-view surveys and their applications for enhancing urban flood simulation. Applying new surveying platforms of UAV, MMS, and USV is given some alternative ways to obtain topographic data from different-source views. Advances in parallel computing and large improvements in urban flood models are further explored.

Chapter 10 concludes the main findings, answering all research questions. Some recommendations are also given.

These chapter guidelines are also represented as an outline diagram in Fig. 1-2.

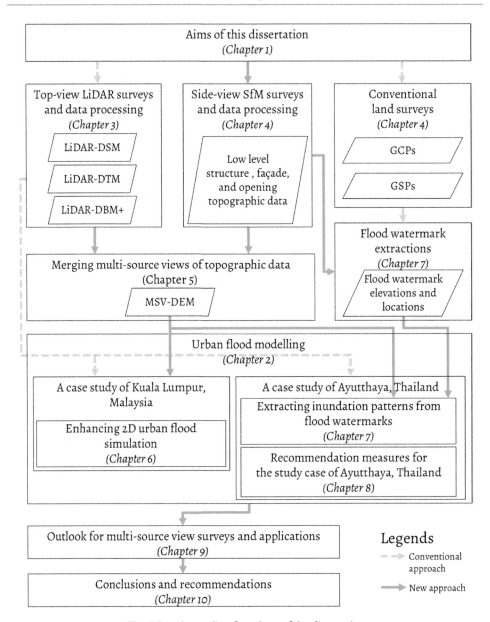

Fig. 1-2.   An outline flowchart of the dissertation

CHAPTER 2
State of the art in urban flood modelling

Urban flooding can happen when discharges exceed riverbank capacities leading to overflows from rivers and reservoirs (fluvial floods); or when intensive rainfalls exceed capacities of drainage systems (pluvial floods); or when massive coastal surges and high tides strike shorelines (coastal floods); or when combinations of these events occur. Hydrodynamic models can play an important role in simulating water level rise, estimating the evolution of inundation extents, and indicating hazard areas of high floods. These simulation results can adequately be used as supporting information for evaluating flood damages from the past, or for predicting floods in the future. An appropriate quality of urban-flood simulation results should replicate and represent key behaviours of flow dynamics as close to reality as possible. In this chapter, approaches to urban flood modelling are introduced in Section 2.1. Principles of 1D urban flood models and quasi 2D-modelling approaches are described in Section 2.2 and 2.3, resp. Principles of 2D urban flood models (Section 2.4) and advances in coupled 1D-2D models (Section 2.5) are then given. Some simulation results of hypothetical cases are compared in Section 2.6. Issues concerning complex urban flood modelling are discussed in Section 2.7.

2.1 Approaches to urban flood modelling

In 2008, more than 50% of the world's population lived in urban areas (UN, 2009). A large number of these urbanising cities have experienced with natural hazards, of which over 70% are flood related incidents (EM-DAT, 2014). Flood hazards are often characterised by magnitudes or intensities, speeds of onset, durations, and areas of an extent (Jha et al., 2011). In urbanising cities, an occurrence of urban floods could be consequences of natural hydro-meteorological phenomenon combined with interactions of both natural catchments and urban structures. Through history, a number of urbanising cities have often been settled in flood-prone areas (e.g. riverbanks, lakesides, and coastlands). When those cities are flooded, floodwaters can interrupt local activities, affecting their daily lives (Fig. 2-1). Threats of these urban flood circumstances can become critical or even catastrophic issues.

Fig. 2-1.   Bangkok residents affected by the worse flood disaster in
Thailand 2011 (source by UN News Centre, 2011)

When floodwaters are entering a city, disturbances, damages, or even losses cannot be avoided, but proactive preventions might help to minimise flood damages and reduce the number of lives lost. While flood protection measures become more important, a balance in structural and non-structural measures crucially needs to be achieved optimal solutions for urban flood managements (Vojinović et al., 2011). Exhaustive understanding of flood hazard consequences for cities needs practical tools for evaluating these flood-protection measures. Physically-based urban flood models as supporting tools are capable of replicating and analysing flood hazards (Schubert & Sanders, 2012). Simulating such floodwater depths are initially used as the main indicator to determine degrees of flood hazards (Vojinović, 2007; Abbott & Vojinović, 2009; Abbott & Vojinović, 2013; Vojinović et al., 2016). Other variables such as flood velocities, flood durations, flood evolution extents, and human activities being interrupted from flooding are also taken into consideration of flood-hazard and flood-damage analyses.

A simple flood model could be used to simulate floods for low to moderate complex areas, and a more complex flood model should be better applied to more complex flood situations like complex urban floods. Generally, one-dimensional (1D) numerical models are often used for simulating floodwaters for open channels, pathways, and drainage pipes. Even though these 1D models may reasonably be appropriate for decision-making, sometimes they may not adequately replicate some complex flood situations of excess floodwaters spilling out of conduits and propagating into floodplains (Djordjević et al., 2004; Mark et al., 2004). For some less complex areas, quasi 2D approaches could be capable of handling such situations. However, a fully two-dimensional (2D) numerical models are more appropriate for simulating flow dynamics in the more complex flood situations (Neal et al., 2009; Hai et al., 2010; Chen et al., 2012a; Chen et al., 2012b; Rychkov et al., 2012; Smith et al., 2012). Moreover, coupled 1D-2D models (Section 2.5) can be further applied to achieve benefits from both 1D and 2D models (Hsu et al., 2000b; Chen et al., 2006). Also, some elaborations are further given hereafter.

2.2 1D flood modelling

Applications of 1D models can provide benefits from their quick and easy modelling setups also their simple and fast predictions. These 1D models have been used in many engineering and planning purposes, for a long time already. In this research, the MIKE 11™ software by DHI™ was chosen as a 1D urban-flood modelling tool. A quality of simulated flood maps is crucially driven by analytical quality of input data. Particularly, key descriptions of surface elevations – topography and submerged surface elevations – bathymetry could be counted as the utmost important input data. When a modelling area have been chosen, topographic data could be the first and foremost important input data for creating1D model schematics, as described hereafter.

2.2.1 Cross sections of river floodplains

Detail descriptions of 1D model schematics are commonly created by using surveyed topographic data. Bathymetries from surveyed cross sections are used for describing and shaping 1D model schematics. Descriptions for each cross section $(XS_1, XS_2, \ldots,$ and $XS_n)$ are defined by a left-bank, a right-bank, and a bottom elevation of canals, rivers, and conduits with a georeferencing location. Cross-sections are specified by values in X and Z coordinates (Fig. 2-2a): XS_{1L} and ZS_{1L} represent the left bank location and elevation, resp; XS_{1R} and ZS_{1R} represent the right bank location and elevation, resp; ZS_{1B} shows the deepest bed elevation at the bottom; and h is a water depth corresponding bed elevation. Typically, all elevation values are referred to the mean sea level (msl) elevation.

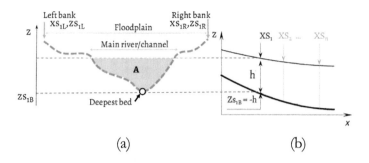

(a)   (b)

Fig. 2-2.   A schematization of (a) a cross section and (b) a longitudinal profile of
a single branch contained several cross sections

From Fig. 2-2b, a ratio between flow area (A) and water depth (h) are assumedly consistent for all identical cross sections. A combination of the cross sections along the same longitudinal profile shapes an open-channel branch (Fig. 2-3), and finally constructs a river network. Owing to this, one open-channel (river) network may contain some branches with a number of cross sections.

Fig. 2-3. A conceptual map of river cross sections in a floodplain
(background by All4Desktop)

2.2.2 Cross sections of urban floodplains

When branch shapes have less complexities (e.g. straight branch with box shape cross sections), addressing only few cross sections may adequately represent such branches (Fig. 2-4).

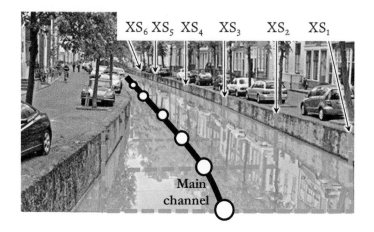

Fig. 2-4. A conceptual map of a canal branch with cross sections in
a city (background by Lezenby, 2013)

Samuels (1990) gave some choices for spacing locations of cross sections that may need to shape natural river geometries suitably. Samuels's research showed that numbers of cross sections not only can define the branch shape but also determine the length of each branch differently. For some examples, two and three cross sections slightly shape the straight branches of 290 m and 328 m lengths, resp. (Fig. 2-5a and b), whereas nine cross sections can better show a curvy branch with 344 m length (Fig. 2-5c).

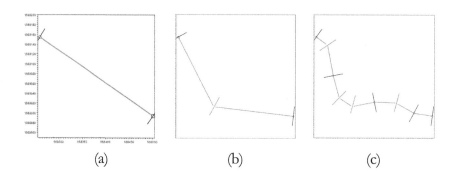

Fig. 2-5. Examples of three different cross-sections spacing: (a) two cross sections,
(b) three cross sections, and (c) nine cross sections

In this section, an example of a 1D hypothetical case was introduced. Nine cross sections were selected for creating a simple branch. A constant width of 40 m was defined for each cross section, which distributedly allocated along this simple branch. For each cross section, key descriptions were defined by elevations of the left and right banks at 3.5 m msl with bottom depth elevation at 0 m msl (Fig. 2-6).

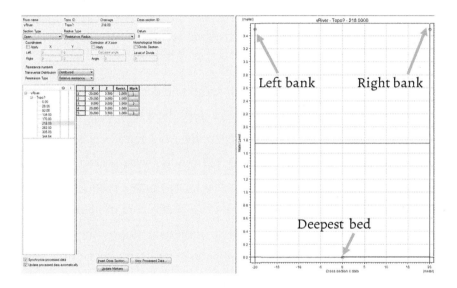

Fig. 2-6.   An example description of a cross section constructed in MIKE 11™

2.2.3 1D schematics of 1D models

All key elements are declared in 1D model schematics for their 1D numerical simulation. Alternating discharge (Q) and water depth (h) nodes are determined in the computational nodes (Fig. 2-7). These computational nodes are automatically generated by user requirements. The Q points are automatically placed midway between neighbouring h points, while h points are located at the defined cross sections, or at equidistant intervals in between.

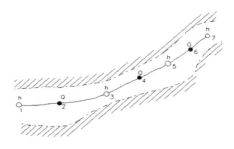

Fig. 2-7.   A schematic of a single branch contained computational grids: h points (white circles) and Q points (black circles), (source by DHI, 2016a)

A distance between nodes in each branch should be declared carefully. When the node spacing is small, it is also necessary to decrease a time step, due to stability conditions of the models are crucially depending on a ratio between a time step and a node spacing. Technically, running 1D models with coarser node spacing (Fig. 2-8a and b) should be faster than smaller node spacing (Fig. 2-8c). However, variability details between two consecutive cross section are more important than the small node spacing. For many mild to moderate slopes, the coarser node spacing could be adequately applied and their coarser-node simulation results still have shown insignificant effects compared to the finer-node simulation results.

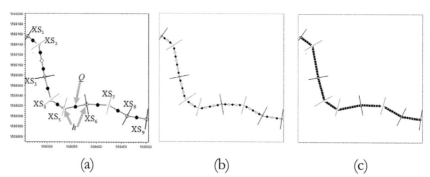

(a)   (b)   (c)

Fig. 2-8.   Example 1D schematics of (a) nine cross sections showed in MIKE 11™, using three different grid spacing: (a) at 50 m, (b) at 20 m, and (c) at 5 m, resp

For 1D hypothetical case, we found that nine cross sections were good enough for shaping a simple branch. This branch and all cross sections were allocated along the same branch in georeferencing system of EPSG 32647: WGS84 UTM 47N. The Q points were automatically placed with coarser node spacing at 50 m between neighbouring h points (at midway when node spacing of two consecutive h points is less than 50 m), while h points were located at defined cross sections (Fig. 2-8a).

2.2.4 Bed resistance conditions for cross sections

Bed resistance conditions are commonly involved in the 1D model schematics (at h points in MIKE 11™). Most forms of friction equations have been developed under normal steady state flows. A friction resistance coefficient is conventionally expressed as Chezy (1776), Manning et al. (1890) friction factors (Eq. 2-1 and 2-2, resp) or other alternatives to estimate the resistance coefficient.

$$v=C\sqrt{Ri}$$

Eq. 2-1

$$v=\frac{1}{n}R^{1/3}S^{1/2}$$

Eq. 2-2

where:

$$C=\frac{R^{1/6}}{n}=MR^{1/6}$$

$$M=\frac{1}{n}$$

C is the Chezy roughness coefficient (m$^{1/2}$ s^{-1})

R is the hydraulic radius (m)

i is the bottom slope (m m^{-1})

S is the slope of hydraulic grade line (m m^{-1})

v is the mean velocity (m s^{-1})

n is the Manning's n coefficient (s m$^{1/3}$)

M is the Manning's M coefficient (m$^{1/3}$ s^{-1})

In this research, the Manning's M coefficient values are mainly used for describing friction forces between two bodies. Manning's M coefficients can also be used as a parameter for energy dissipation. The coefficient of friction dominantly depends on surface topography (bathymetry) materials and their coverage, following the values found by Chow (1959).

2.2.5 1D De Saint-Venant flow equations

Physical laws are mandatory needed for understanding the entire process of complex flood dynamics which replicated in numerical model simulation. A conservation law states that a quantity element in an isolated system's evolution remains constant, and it is an invariant at all time, though it may change form. It is useful approximate solutions in calculations that can be corrected by finding the nearest state satisfied the suitable conservation laws. With respect to hydrodynamic physics, three principles that describe fluid flow, mass transfer, and heat transfer derive from the three well-known conservation laws of classical physics namely; (a) conservation of mass (law of conservation of mass); (b) conservation of momentum (Newton's second law of motion); (c) conservation of energy (first law of thermodynamics). Wherever flow remains nearly horizontal, varying smoothly at a point in a continuum, the mass-momentum and mass-energy couples are equivalent concepts. However, when a discontinuity (e.g. hydraulic jumps and bores) appears, these two concepts may not be equivalent, and their equations will produce different answers. Even though the mass-momentum couple of conservation laws is applicable to both continuous and discontinuous situations, the mass-energy may not be appropriate. Therefore, only the mass-momentum couple is further discussed and physical principles in approximate solutions can be translated into equations. A derivation of these laws follows a concept of a fix control volume used an Eulerian view of motion and represented in a Cartesian coordinate system.

In 1D models, the governing equations describing changes in flow velocities and water depths are based on the conservation principles of mass and momentum. They are referred as the De Saint-Venant (1871) Equations, which can be written in different forms and different approximations. For example, the discharge flow form of the continuity equation is emphasised as:

$$\frac{\partial A}{\partial t} + \frac{\partial Q}{\partial x} = 0$$

Eq. 2-3

In addition, the moment equations can be shown as:

$$\underbrace{\frac{\partial Q}{\partial t}}_{(i)} + \underbrace{\frac{\partial}{\partial x}\left(\frac{Q^2}{A}\right)}_{(ii)} + \underbrace{gA\frac{\partial h}{\partial x}}_{(iii)} + \underbrace{\frac{g}{A}\left(\frac{Q|Q|}{C^2 R}\right)}_{(iv)} = 0$$

Eq. 2-4

where:

A is the flow area (m²)

t is the time (s)

Q is the discharge (m³ s⁻¹)

x is the grid size in the x direction (m)

g is the gravitational acceleration (m s⁻²)

h is the water depth (m)

C is the Chezy friction factor (m¹ᐟ² s⁻¹)

R is the hydraulic radius (m)

From Eq. 2-3, the equation combines with the four terms: (i) a local momentum term; (ii) a convective momentum term; (iii) a gravity term; (iv) a friction term. In 1D model, flows are commonly simulated in the defined conduits (e.g. open channels and pathway channels). Therefore, these 1D equations are adequately used for less-complex flood analyses, when the flows are dominantly driven in one-dimension (e.g. along x direction in horizontal plain) and average slopes of channel beds should be relatively small. According to this assumption, velocities are uniform over cross sections only in single direction (x), but they are negligible in both lateral (y), and vertical (z) directions. They have no lateral flows and vertical accelerations are neglected. The water depths across all sections are averaged in horizontal, and their pressures are hydrostatic. The effects of boundary friction and turbulence are also taken into consideration through resistance laws. Adopting these De Saint-Venant Equations could be later solved in numerical solutions (examples in Abbott, 1980; Cunge et al., 1980).

2.2.6 Boundary conditions for 1D models

External boundary conditions are required at all 1D model boundaries (i.e. all upstream and downstream ends of each branch). If topographic data are the most important input data, then the boundary conditions could be counted as the second one. In MIKE 11™, determining and locating each boundary condition should be located at a sufficient distance to be sure that they are not intervened by other boundaries, and they cannot be assigned at a connected junction. These boundary conditions are defined as constant values of h (water depth) or Q (discharge); time-series values of h or Q; or a relationship between h and Q (e.g. a rating curve). The choice of boundary condition depends on physical situations being simulated and availability of input data (Fig. 2-9). Typically, time-series discharges should be used for upstream boundaries, and time series of water level or a reliable rating curve should be applied for downstream boundaries.

In time-series, these boundary data could be obtained from different measuring sources (e.g. water depths and discharges obtaining by systemic gauges). Sometimes the time-series data can also be achieved by some simulation results, which transferred from other hydrological or hydraulic model predictions. When such transferred results should be simulated in larger area for the entire model area (domain), they can apply as input for smaller area located in the same domain. When an interval of time-series data is greater than a time step used in the simulation, applying linear interpolation are also capable of creating the intermediate values for the boundary conditions. These pre-processing data interpolations should be applied only for smoother transitions between the known values. It is important to establish the appropriate boundary conditions for the solution space. The better boundary conditions could have less simulation instability issues.

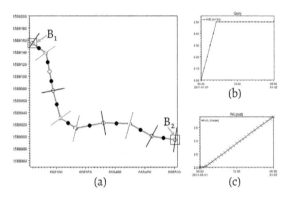

Fig. 2-9.   Examples of upstream and downstream boundaries (a) at locations 'B1' and 'B2', resp: (b) time-series values of Q (discharge) at B1 and (c) time-series values of h (water depth) at B2

2.2.7 Initial conditions for 1D models

While boundary conditions as hydrological input data are used for the entire period of simulation times, initial condition values use only once at the initial time ($t = 0$). The initial h (water depth) and Q (discharges) can be determined by either users or automatic start. For automatic start in MIKE 11™, initial conditions are determined by using values at the initial time of given boundary conditions. It is possible to use earlier simulation results as input for initial conditions, aka hot-start. This hot-start approach can be used only when earlier and current simulation domains are both compatible.

2.2.8 Sample 1D simulated results

In 1D models, numerical solutions of MIKE 11™ are based on hydrodynamic governing equations. These solutions of continuity and momentum equations can be calculated on an implicit finite difference scheme using 6-point Abbott's scheme developed by (Abbott & Ionescu, 1967). An overviewing simulation result can represent in different ways. For example, a 1D simulated result was represented along a longitudinal profile of the selected branch (Fig. 2-10a).

When a longitudinal branch has been selected, 1D simulation results can be visualised, animated, and synchronised with the selected branch (location 'LP' in Fig. 2-10a; Fig. 2-10c) and the selected cross section (location 'XS$_6$' in Fig. 2-10a). During a simulation, when a simulated water level rises above the maximum elevation of the cross section (i.e. left bank or right bank elevations) a hydraulic area is further calculated by assuming the river banks extend vertically upward (Fig. 2-10b). The simulation will be terminated when the simulated water levels become more than four times (by default) higher than the bank elevations.

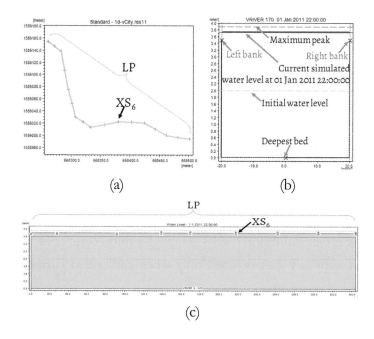

Fig. 2-10. Examples of 1D simulated results: (a) an outline of the 1D schematic; (b) a water level at the observed XS$_6$ cross section; (c) a water level along a longitudinal profile (LP)

2.3 Quasi 2D approaches from 1D models

Conventional 1D flood models are often used for simulating less-complex flood situations. Many modellers and practitioners are still favour using these simple 1D models, due to these simple models are relatively easy to setup and calibrate. Their 1D simulation results are reasonably explained to decision makers and communities for many situations. However, simulated flows using these conventional 1D models may not be plausible for more-complex flood situations.

2.3.1 Quasi 2D approaches to river floodplains

One limitation of conventional 1D models is that they simply assume all flow moves along one dimension. For given cross sections of a river, all of flows are assumed to move either downstream or upstream, along with such single direction (e.g. x direction). They can only use for replicating one water surface elevation and one total flow at a given time step for each given cross section. However, applying quasi 2D approaches in 1D models (examples in Fig. 2-11) for creating more complex 1D model schematics could be feasible.

Whenever a given topographical data are available for the entire area, creating wider cross-sections for both channel and floodplain areas can be made. For example of a given floodplain, natural levees are formed along the river due to the abrupt reduction in flow velocity of the entering water. In this situation, rivers and floodplains may be schematized separately with several different types of 1D model schematics depending on topography of flood plains and flow natures.

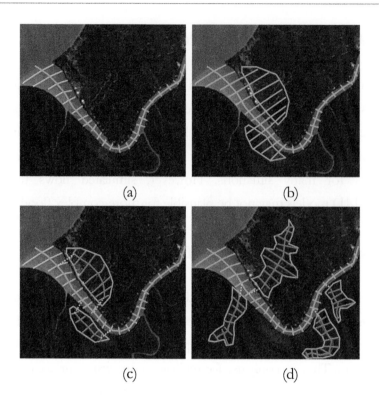

Fig. 2-11. An example of (a) a conventional 1D channel schematic and three examples 1D model schematics applying different quasi 2D approaches: (b) a widening cross-sections scheme; (c) a parallelising branch scheme; (d) a tributary branch scheme

2.3.2 Quasi 2D approaches to urban floodplains

These quasi 2D approaches are capable of replicating dynamics flows for not only river floodplains but also urban floodplains. While existing branches can also be included from new branches (Fig. 2-11d), such tributary branches can give more correct routeing (delay and attenuation) of the hydrograph. Parallelising and tributary branches (Fig. 2-11c) are often used in quasi 2D approaches. Lhomme et al. (2006) also, noted that a width of an extended cross-section (Fig. 2-11b) in the floodplain should not be larger than three times width of main channel (Fig. 2-11a).

For 1D hypothetical case, extending cross sections were created at three times of the normal branch widths (examples in Fig. 2-12).

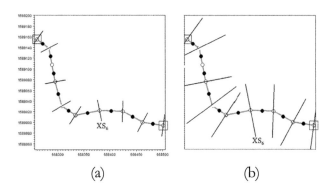

(a) (b)

Fig. 2-12. Two different examples of cross-sections widening: (a) cross sections concerning the channel width, (b) cross sections concerning floodplain width

The normal width description of channel cross sections (XS_1, XS_2, …, and XS_n) can be defined by the left-channel bank, right-channel bank, and bottom depths with geolocations (location 'XS$_6$' in Fig. 2-12a; Fig. 2-6). Whereas the wider width descriptions of the floodplain cross sections can be defined by the left floodplain and right floodplain banks (location 'XS$_6$' in Fig. 2-12b; Fig. 2-13).

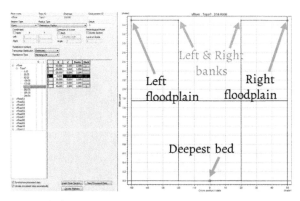

Fig. 2-13. Example descriptions of a widening cross section in MIKE 11™

For some less-complex urban areas, floodwaters can exceed drainage capacities behaved as floodplain areas, especially in a conjunction area between an open channel and a low-lying area. Widening cross sections (location 'B' in Fig. 2-14) from normal widths of the existing canal (location 'A' in Fig. 2-14) should be adequate to simulate the exceedance floodwater over mimetic floodplains (the widened or extended cross sections over urban landscapes) in 1D models.

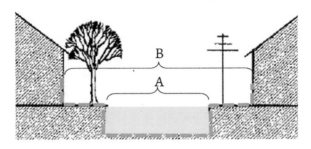

Fig. 2-14. A conceptual map of two cross-sectional profiles: (A) an existing cross section of an open channel; (B) a widening cross section concerning urban floodplains

In urban areas, flood flowing through surface pathways can also behave as slightly the same as flows in a channel. An artificial (surface) pathway could be generated along street, road, and alley paths. Such artificial pathways can also be feasible to apply quasi 2D approaches from conventional 1D models. As almost the same as a river network, a surface drainage network may shape with many surface pathways with a number of pathway cross sections (Fig. 2-15).

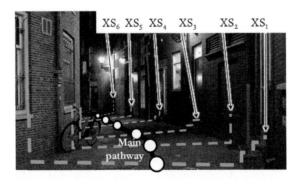

Fig. 2-15. A conceptual map of pathway cross sections in urban landscapes, (background by PsychaSec, 2013)

For more complex of 1D hypothetical case, channel cross sections (Fig. 2-16a) can also be determined using the same criteria as given in Section 2.2.1. Despite only widening channel sections, many pathway cross sections were also shaped. Descriptions of such pathway sections were defined using the left, right, and bottom depths of surface elevations of the road including stairs and kerbs, positioning in the same georeferencing systems. The series of these sections were then gathered in order to shape all channel and pathway branches (Fig. 2-16b).

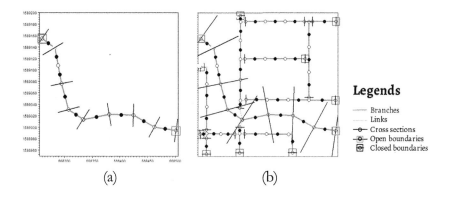

(a)　　　　　　　　　　　(b)

Fig. 2-16. Two examples of 1D model schematics; (a) concerning only the channel width; (b) concerning a channel, floodplains, and surface pathways

2.4　2D flood modelling

Whenever simulating multi-flow directions is of interest and 1D models no longer capable for such complex flood situations, 2D and 3D models will be suitable for simulating such flood dynamics in multi-flow directions. Even though the closest complex flood situations could be replicated in 3D flow models, simulating these flows in 2D models seem to be more practical and straightforward for most cases of urban flood predictions. Also, simulating in 2D models is technically faster than many sophisticated 3D models. In general, 2D models simulate flows on horizontal plain (x and y directions). These horizon-flow directions can be larger than a vertical-flow direction (z directions). Therefore, such vertical flows could be neglected or considered as a depth-averaged form. An assumption of these (nearly) horizontal flows is commonly indicated by terms of shallow-water equations, which allow simplifications in mathematical formulations and numerical solutions. In this research, MIKE 21™ software by DHI™ was chosen as 2D urban flood modelling tools.

2.4.1 2D schematics of 2D models

Schematics of 2D urban flood models are commonly created by using surveyed topographic data of a city. In 2D numerical solutions, floodwater flows are simulated on computational grids of 2D model schematics. These 2D schemes contain values of topographic data, initial, and boundary conditions for the entire model area (domain) and they are commonly created in the same resolution. Structured and unstructured grids are typically two main distinct 2D schematizations. In this research, we mainly focused on the former structured grid, which appears more convenient and most high-resolution topographic data are widely provided in square structured grids.

Emerging remote sensing technologies and their huge improvements could acquire high details of topographic data of a city. Such high-resolution topographic data are becoming required, especially for the complex cities. Amongst of these topographic data, square-structured grids of digital elevation models (DEMs) are commonly used for representing raster-based topographic data, which could be obtained from different remote sensing sources (e.g. satellite data and images, aerial photos, LiDAR data). The straightforward on these square DEM grids is consequently easier to code in data processing steps (Peaceman & Rachford, 1955). Such DEMs can be effectively represented and easily translated for creating 2D model schematics (examples in Fig. 2-17).

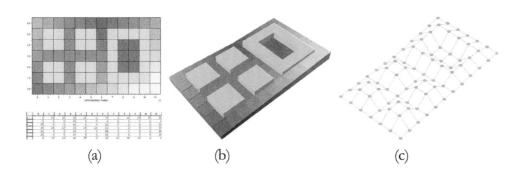

(a)   (b)   (c)

Fig. 2-17. An example of a DEM: (a) a square-structured of DEM (up) and their tabular values (below). Two examples of different representations of 2D schematics: (b) a perspective view of square-structured grids; (c) a perspective view of connecting nodes

Qualities of urban-flood simulation results are crucially depended on availability of these topographic data. The more details of complex urban topographies should better describe more key components in 2D mode schematics. Technically, such complex topographies are commonly delineated into a simplified 2D model scheme as 2D computational grid cells. When these computational-grid resolutions are increased (grid sizes are decreased and a space of each computational grid approaching closely to zero), the details of such 2D schemes will also be improved (Wright, 2005). However, it should be more practical to delineate real complex city's topography and translate their key components into a simplified 2D scheme.

In this research, the structured square grids were used as the main computational grid system for the 2D model schematics. All topographic descriptions were analysed, simplified, and next translated for matching with grid-based resolution of the 2D model schematics (Fig. 2-18). When topography variables were determined to these computational grids, other key conditions (e.g. bed resistance conditions) will then be defined into the same grids. All boundary and initial conditions were also determined and later used for simulating their flood dynamics based on principle of hydrodynamic equations.

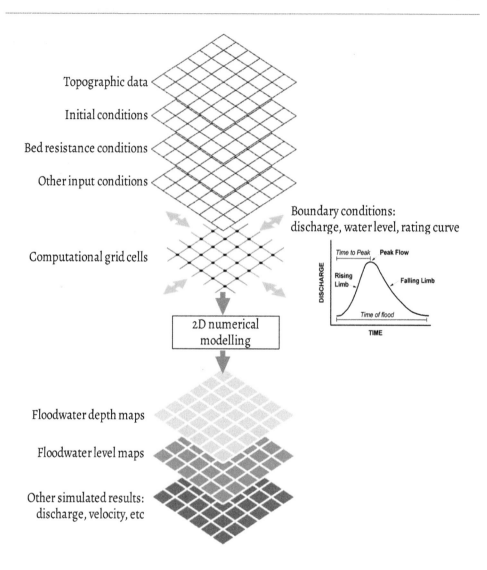

Topographic data

Initial conditions

Bed resistance conditions

Other input conditions

Computational grid cells

Boundary conditions:
discharge, water level, rating curve

2D numerical
modelling

Floodwater depth maps

Floodwater level maps

Other simulated results:
discharge, velocity, etc

Fig. 2-18. A schematization of initial and boundary conditions in
2D numerical model setup

For 2D hypothetical case (Fig. 2-19), 2D model schematics were created from
different types of DEMs. All these 2D schematics were created at 5 m grid
resolution with 50 x 50 square grid cells in the same georeferencing location of the
1D hypothetical case (EPSG 32647: WGS84 UTM 47N).

Fig. 2-19. An example of 2D model schematic created in 5 m grid resolution. Terrains and channel bathymetries are showed in grey boxes urban structures are showed in other colour boxes

2.4.2 Bed resistance conditions for 2D models

Bed resistance conditions (i.e. Manning's M friction coefficient values) are typically defined on computational grids of 2D model schematics. In 2D flood predations, applying such resistance values to steep slope channels are more significant than applying them to floodplains. Achieving such specific friction values to floodplain for each specific surface (in laboratory tests) are varies and could be hard to verify such values for the mix and complex surfaces in reality. Even if such friction phenomena on different surfaces (roads and so on) and vegetation are known, localising such resistance values on fine-floodplain 2D schematics could be intensive works. When floodwater levels have slowly changed, applying such resistance values in rough-floodplain 2D schematics slightly show more effects than applying them in the fine-floodplain 2D schematics. Then uniformly distributing friction values over floodplain 2D schematics apparently sounds more practical (Pappenberger et al., 2005). Therefore, parameterizing such resistance values uniformly for the entire fine-floodplain 2D scheme was applied in this research.

2.4.3 The 2D De Saint-Venant flow equations

The two-dimensional (2D) Saint-Venant equations (aka shallow water equations) are derived from the principles of mass and momentum conservations. The depth in shallow-water flows should be small when compared to the wavelength of the disturbance. Since a gradually varying free surface is assumed as the flow of an incompressible fluid, such that vertical accelerations (the gravity acceleration) are negligible.

The 2D Saint-Venant equations could be written in different ways, they also have different forms and different approximations. For example (McCowan et al., 2001), when the shallow-water flows are nearly horizontal, it allows a considerable simplification by assuming the pressure distribution to be hydrostatic. In 2D horizontal dimensions (in Cartesian coordinate system), depth-averaged flows form of the continuity equation can be obtained in the following form:

$$\frac{\partial h}{\partial t} + \frac{\partial (uh)}{\partial x} + \frac{\partial (vh)}{\partial y} = 0 \qquad \text{Eq. 2-5}$$

with the momentum equations expressed as:

$$\frac{\partial u}{\partial t} + u\frac{\partial u}{\partial x} + v\frac{\partial u}{\partial y} + g\frac{\partial h}{\partial x} + gu\frac{\sqrt{u^2 + v^2}}{C^2 h} = 0 \qquad \text{Eq. 2-6}$$

$$\underbrace{\frac{\partial v}{\partial t}}_{(i)} + \underbrace{u\frac{\partial v}{\partial x} + v\frac{\partial v}{\partial y}}_{(ii)} + \underbrace{g\frac{\partial h}{\partial y}}_{(iii)} + \underbrace{gv\frac{\sqrt{u^2 + v^2}}{C^2 h}}_{(iv)} = 0$$

where: u, v are depth-averaged velocities in x and y directions (m s^{-1})

From Eq. 2-6, four main terms represent (i) the local momentum term, (ii) the convective momentum term, (iii) the gravity term, (iv) the friction term. In a very large water surface as lakes and oceans, all four terms are commonly used including additional forces, e.g. wind shear stresses and Coriolis forces. In urban flooding, (ii) the convective term and other addition-external forces can be considered negligible for increasing model stabilities.

2.4.4 Boundary conditions for 2D models

External boundary conditions are required at all model boundaries, as almost the same as in boundaries for 1D models. In 2D models, the MIKE 21™ solves the partial differential equations that govern nearly horizontal flows. The variables of flux densities and water surface elevations in the x-direction and y-direction must be specified in all grid cells along the open boundary at each time step. The choice of boundary conditions between the flux and the surface elevation must be chosen for determining at such open boundaries on square grid cells. Actual values of water levels or fluxes at each boundary are specified in one of five different formats: constant value, sine series, time series, line series, transfer data, and rating curve.

When determining fluxes at boundaries, it could be either used discharges or fluxes through the boundary and the flow direction. Whereas choosing water surface elevations (water levels), it could be constant or vary along the boundary line. When selecting time-series boundary variations, boundary values can be applied for multiple open boundary cells. When results of other simulations are available, such transferred simulation results are capable of applying as input for such boundaries. Modellers can also have a degree of freedom to define their open boundary positions, manually. Moreover, flow directions should be defined. Accurate flow directions are important when the flow is moving into the model (inflow), while they are of less importance at the outflow. Because any error at the inflow boundaries could be transported into the model and such errors may, therefore, cause model instabilities.

2.4.5 Initial conditions for 2D models

The initial water depth (h) or surface elevations must be provided in the 2D numerical model setups, which can be specified in two ways: as constant values for the respective area; or as spatially distributed values over the entire domain. For automatic start in MIKE 21™, initial conditions are determined by using values at the initial time of given boundary conditions. It is possible to use earlier simulation results for determining initial conditions, aka hot-start. This hot-start approach can be used only when the earlier and new simulation domains are both compatible.

Because the model initializes fluxes or current velocities to zero, an initial surface elevation must be specified in agreement with these conditions. In practice, this means that the initialise values should agree with the boundary conditions at the first time step. For example, when starting simulation at high-water levels with a boundary value of 2 m msl, the initial surface elevation should also specify to ca. 2 m msl (Fig. 2-20). When a model domain is large and water surface levels at the open boundaries have substantial differences, averaging water surface elevations at open boundaries should be applied rather than customising initial surface elevations at each grid point.

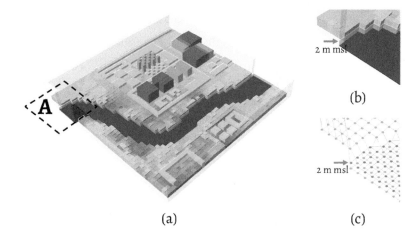

(a)   (c)

Fig. 2-20. A perspective-view example of (a) 2D schematics with initial water surface level at 2 m msl. Two zoomed-in examples of location 'A' visualised in (b) squared grids and (c) connecting nodes

2.4.6 Sample 2D simulated results

The 2D flood maps using MIKE 21™ are simulated based on square structure grid cells and constructed in the space of grid point results and these 2D simulated maps can show in different representations. For 2D hypothetical case, an example of 2D simulated floodwater depths was visualised in square grid cells (Fig. 2-21).

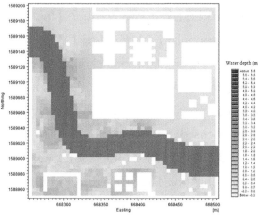

Fig. 2-21. An example of a 2D simulated floodwater depth map

The volumetric flux vectors with different amplitudes and directions (Fig. 2-22a) including floodwater depths (Fig. 2-22b) can show in the square grid cells.

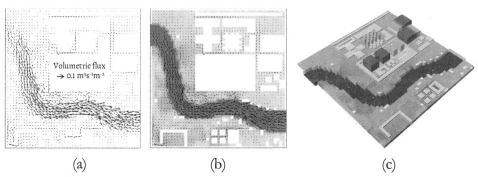

(a)   (b)   (c)

Fig. 2-22. Examples of 2D simulated flood maps: (a) amplitudes and directions of flood-wave propagations; (b) floodwater levels with amplitudes and directions of flood-wave propagations; (c) square grids of floodwater levels over topography in a perspective-view

Moreover, a perspective view of squared grid visualisation of flood inundation maps was also given (Fig. 2-22c).

2.5 Coupled 1D-2D modelling

Advances in further developments of coupling 1D and 2D models, aka a coupled 1D-2D models, should also be taken into account. Nowadays, coupled 1D-2D models are becoming more popular for both pluvial and fluvial flood applications. These coupled 1D-2D models are appropriately used to simulate urban floods, especially in the conjunction areas between confined conduits and floodplains. Where some studies include possibilities to link 1D subsurface networks over 2D floodplains, we mainly focused on coupling 1D-channel networks over 2D floodplains. In this research, MIKE FLOOD™ software by DHI™ was chosen as a coupled 1D-2D modelling tool. The MIKE FLOOD™ are capable for linking 1D models (MIKE 11™) with 2D models (MIKE 21™) in the same software package.

In coupled 1D-2D models, when floodwaters start to exceed conduit capacities from 1D models, they can freely flow out of conduits and propagate into lowland areas of floodplains in 2D models. These excess floodwaters rather flow naturally on terrains and they are commonly diverted by urban features above those terrains. When a simple 1D model is coupled with a fully 2D model (Fig. 2-23), their coupled 1D-2D model simulation results are typically capable of replicating exceedance floodwater flows between 1D-channels and 2D floodplains (Neal et al., 2009; Hai et al., 2010; Chen et al., 2012a; Chen et al., 2012b; Rychkov et al., 2012; Seyoum et al., 2012; Smith et al., 2012).

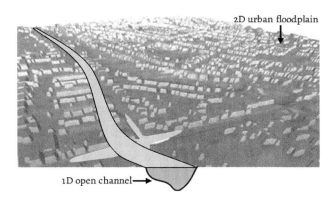

Fig. 2-23. A conceptual map of a coupled 1D-2D modelling for
a 1D open-channel network and a 2D urban floodplain

When these coupled 1D-2D models have emerged, they are become more widely used to simulate urban floods, which commonly found in pluvial and fluvial flood simulations, especially for complex cities (Vojinović & Tutulic, 2009). In coupled 1D-2D simulations, both conduit networks in 1D models and floodplains of 2D models are taken into account (Stelling et al., 1998). Floodwater flows spilt from an open and/or pathway channels of 1D models can reasonably be further simulated and synchronised with complex floodplain areas on 2D models (Fig. 2-24).

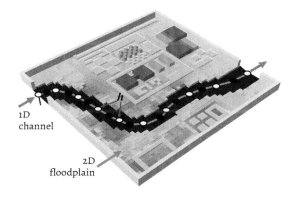

Fig. 2-24. An example of a 1D-channel scheme coupled with
the square grids of a 2D floodplain scheme in a perspective-view

The MIKE FLOOD™ software is suitable for coupled model simulation for both a 1D-1D and a 1D-2D coupling models. Using a coupled 1D-2D approach enables best features of both 1D and 2D models to be utilised, while at the same time compromising limitations of different resolutions for each model. In 2D models, square grid schematics were chosen for this research. These structured grid geometries are subdivided into a series of internal structures with a cell-to-cell method. Each internal structure has a bed level and a width determined by resolutions of each structured grid point. In this research, we mainly focused on coupled 1D-2D models, which can (almost) simultaneously replicate floodwaters in both 1D channels and 2D floodplains and lateral links were applied for internal structure links. During simulations, water levels at each h point from 1D schemes using MIKE 11™ and 2D grid cells using MIKE 21™ were assigned to specific internal structure locations (Fig. 2-25). Technically, these lateral links allow a string of MIKE 21™ grid cells to be laterally linked to each h point of MIKE 11™.

Floodwater flows through this lateral link were calculated using a structure equation as the same equation as used in MIKE 11™. Exchange flows are typically modelled by using broad crested weir equations or depth-discharge curves based on water level differences (Lin et al., 2006; Evans et al., 2007; DHI, 2016b). Flows from each internal structure are then distributed to/from h points and square grid cells. All of information available in this internal structure geometry were then utilised. Structure points are then defined in each grid cell. Bed levels from each grid cell are next used as structure bed levels. Calculations of widths, bed levels, and interpolated water levels, should be determined at existing calculation points onto these internal structures. When h points lie within a range of influence of internal structures, flows should be distributed across those points according to water depths at each point. When none h point lies within ranges of influence, flows could be distributed to the nearest upstream and downstream points by using a distance-dependent interpolation. The same approach could also be applied to square grid cells.

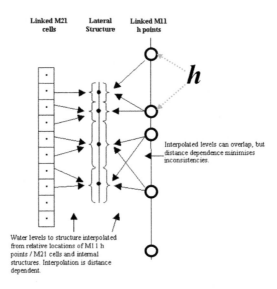

Fig. 2-25. A conceptual map of lateral link (source by DHI, 2016c)

A limitation to this lateral-link solution is that a complicated momentum exchange processes that characterise floodplain boundaries are not modelled (due to the fact that these processes intimately depend on complex 3D flow patterns in the river, which by their definition cannot be resolved in 1D river models). Progress towards an improved model representation is given in (Liang et al., 2007). However, simulation processes in the case of fully 3D modelling are rather complex and it can become computationally very expensive (i.e. time consumption). In this research, usefulness of coupled 1D-2D models seems to be promising of replicating complex urban floods with reasonable computational costs.

In this section, an example of a coupled 1D-2D hypothetical case was created. Again, the Manning's M value of 40 was uniformly applied for all cross sections of the 1D model schematic, and Manning's M value of 30 was entirely applied to floodplain area of the 2D model schematic. The initial water level of 2 m msl was globally assigned for both 1D and 2D models (an example of water level initial of 2D model in Fig. 2-20). For the 1D model, upstream boundary conditions used a time-series values of Q (discharges) at B1 (location 'B1' in Fig. 2-9a; an example in Fig. 2-9b), and downstream boundary conditions used a time-series values of h (water depths) at B2 (location 'B2' in Fig. 2-9a; an example in Fig. 2-9c). For the 2D model, upstream and downstream boundary conditions were blocked out to avoid repeat value assignments at the boundaries. For this coupled 1D-2D models, when the 1D and 2D models were carefully created in the same georeferencing system of EPSG 32647: WGS84 UTM 47N, they were feasible for coupling both models and the lateral links as internal structures were applied to this coupled 1D-2D model (Fig. 2-26).

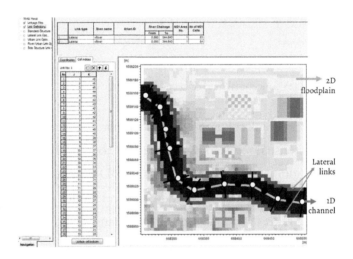

Fig. 2-26. An example of a coupled 1D-2D modelling in DHI MIKE-FLOOD package

2.6 Comparisons of simulated results

2.6.1 Calibration and validation basis

An appropriate quality of flood simulation results crucially depends on essences of input data. Typically, these results should replicate behaviours of flooding as close to observed measurements as possible. To achieve the best fit of the simulated results, a calibration as an iterative procedure in physically based models often uses parameter values for adjustments. Indeed, they should not be identically matched and the calibrations should, therefore, be handled with care. When calibrations are involved, verifying these models are needed for evaluating such simulated results continue to be reasonable for another set of observed measurements.

Parameterization in modelling is usually applied the Manning's friction coefficients for more adjustments in calibrating processes. The models should be parameterised using engineering judgement informed by experience. Such issue, however, is still debated in the literature (Beven, 2000; Cunge, 2003). For example, the parameterization of bed friction is a much more important issue, because flow predictions (velocity and flood wave celerity) could be crucially depending on friction parameter values. Over adjusting values far above or far below normal ranges (Chow's criteria) can lead falling into the trap of force-fitted model.

To avoid unreasonably or over adjusting parameters, the calibration should be handled with care. For certain parameters, adjusting and limiting the values to their normal ranges can avoid the force-fitted model. These certain parameter values are achieved by analysing and run experimental testing in a laboratory, or such values could be quantitatively defined as descriptions. Some guidelines may be given in a range of appropriate values, but the precise value is still unknown. Exceeding these limitations may lead incorrect-simulated results. Hence, it is important that the modeller learns to live with the limitations of the model, and to avoid of force fitting the model by adjusting parameters outside their normal ranges of values.

Validation of physically-based models is a crucial process to evaluate such simulated results continue to be reasonable when compared to another set of observed data (as benchmark). Traditionally, such observed data using in validation process could be either water level or discharge time-series data (Gee et al., 1990;

Bates et al., 1992), which commonly recorded by systemic gauges in the main channel networks. It may be more difficult to find sought of these gauges distributedly installed to keep records the overland flow on land.

Even though flood extent maps could be achieved by using aerial photos (Connell et al., 2001; Overton, 2005; Yu & Lane, 2006), airborne and satellite Synthetic Aperture Radar (SAR) data (Horritt, 2000; Brivio et al., 2002; Rosenqvist et al., 2002; Townsend & Foster, 2002; Bates et al., 2006) and post-event LiDAR survey of flood deposits (Lane et al., 2003) and such maps could represent the extent of inundated areas, they may be inconsistent with the peak time of flooding. Flood watermarks could be capable of representing these flood peaks. Some research was further explored benefits of using post-flood analyses for extracting such missing flood peaks from watermarks (Chapter 7).

2.6.2 Comparisons of simulated results for 1D hypothetical case

The MIKE 11™ packages are typically used for simulating 1D models with or without adopting quasi 2D approaches. When a simulated water level rises above the maximum elevation of each cross section, i.e. left bank or right bank elevations, the hydraulic area can still be calculated by extending river banks upward.

For the 1D hypothetical case, at the location 'XS$_6$', 1D simulated results showed that current floodwaters were raised above the river banks (3.5 m msl) at the same water level (3.75 m msl) for both the normal width section (Fig. 2-27a) and the widening width section (Fig. 2-27b).

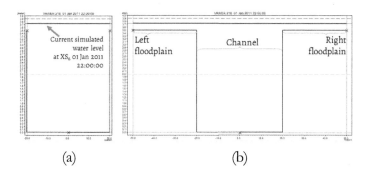

(a)   (b)

Fig. 2-27. Examples of the coupled 1D-2D simulation results at location 'XS₆' when applying different cross-sections widths: (a) the normal cross sections and (b) the widening cross-sections

This MIKE 11™ software is capable for providing conventional 1D flood maps (examples in Section 2.2.8) also (approximate) 2D flood maps. These approximate 2D flood maps are commonly represented in square grid cells and constructed through interpolation into the grid spacing results. Even though conventional 1D simulated results showed the same water levels for both the normal cross sections and the widening cross-sections, their approximate 2D maps represented different sizes of flood inundation areas (examples in Fig. 2-28).

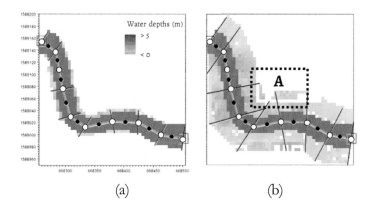

(a)   (b)

Fig. 2-28. Two examples of approximate 2D flood depth maps simulated in 1D hypothetical case when applying different cross-sections widths: (a) only concerning channel widths (b) concerning floodplain widths (adopting quasi 2D approaches)

Due to 1D models simply assume all flows to move either downstream or upstream, along singular dimension of widened cross sections. The consequence of this is that there are only one water depth and one total flow at a given time step for the given cross section. When the widening cross sections are not well constructed, adopting such 1D models for floodplain simulation could have overestimated results (Hervouet & Janin, 1994; Hestholm & Ruud, 1994; Hervouet & Van-Haren, 1996). Too much widening cross sections may result in incorrect flood propagations (location 'A' in Fig. 2-28b). Due to high walls are surrounding the area 'A'. Therefore, it should not have any floodwater entering this area.

In MIKE 11™ model, branches could be determined as open channel and surface pathway descriptions. It is relatively easy to include new model branches to existing networks. When proper quality of topography data is available for a complex city, it is, therefore, clear where a new branch should be defined into the model (Fig. 2-29). These branches commonly route under gravitational conditions. The connection between different branches may be included, especially in larger networks for either computational balancing or bypassing branches.

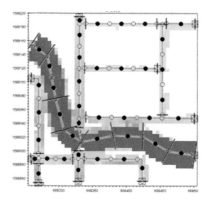

Fig. 2-29. An example of approximate 2D flood-depth map simulated by quasi 2D approach from the 1D model: cross sections concerning the channel and pathway descriptions

Cross selections of model branches are typically based on a compromise between computational time and level of detail. It should be advisable to limit a number of branches in an initial setup and only include more branches, if required. A carefully selected 1D computational grid will avoid many problems during the calibration and application phase.

Even though these quasi 2D approaches can cope with complex pathways and provide some reasonable simulation results, limitations from their 1D flow simulation still cannot provide reasonable multi-dimensional flows for more complex flood situations (Goodell, 2013). Moreover, adopting 1D modelling schemes is hard to give detailed guidelines for how to design the computational grid for all cases. Manually editing detail descriptions for each channel and/or pathway is still time consuming and a lot of labour-intensive efforts, especially a complex city.

2.6.3 Comparisons of simulated results for 1D versus 2D hypothetical cases

A 1D numerical modelling solution can reliably be utilised to simulate the results, which is safe for decision-making. Many researchers and practitioners still use 1D models over other approaches. Some of the commonly perceived reasons are that these models are relatively easy to set up, calibrate, and explain. However, in 1D numerical solutions, the flows assume to be a unidirectional flow. It is clear that the flows are confined into their open conduits, e.g. open channels and pathways (Fig. 2-30). Even though complex-tributary branches of quasi 2D approaches from 1D models can give more correct routeing, surcharge floodwater flows should be applicable only when the cross sections are well constructed. 1D simulated results are dominantly concerned in x direction, yet it cannot truly replicate lateral flows.

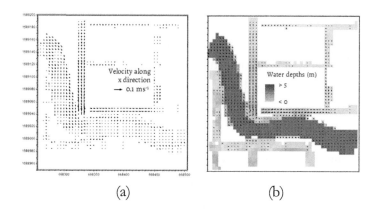

Fig. 2-30. Two examples of approximate 2D flood maps simulated by using 1D model:
(a) amplitudes and directions of flood-wave propagations; (b) floodwater depths with
amplitudes and directions of flood-wave propagations

Whereas the general characteristic of full 2D flows is sufficient to consider as a
bidirectional flow which is concerned a horizontal (x and y) direction with a depth-
averaged form (averaged in z direction). Therefore, the floodwaters propagated
over the floodplains in 2D simulated results appear to show more reasonable than
the complex-tributary branches in 1D models (Fig. 2-31).

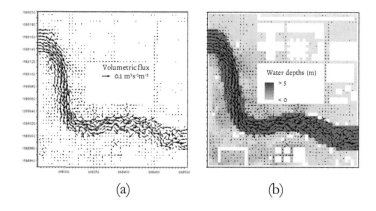

Fig. 2-31. Two examples of 2D flood maps simulated by using 2D model: (a) amplitudes and
directions of flood-wave propagations; (b) floodwater depths with amplitudes and
directions of flood-wave propagations

Moreover, an assumption of these nearly horizontal flows in 2D models is commonly indicated by terms of shallow-water equations, which allow a considerable simplification in mathematical formulations and numerical solutions. Even this assumption not true flows as in the 3D analyses, but its 2D simplification can yield reasonable results to replicate flood behaviours for complex cities.

2.6.4 Comparisons of simulated results for 2D versus coupled 1D-2D hypothetical cases

Some recent research investigates in the coupled 1D-2D modelling at the level of accuracy and efficiency details (Morales-Hernández et al., 2016). In the coupled 1D-2D simulated results of the hypothetical case (Fig. 2-32), the floodwater extents and flows were well captured compared to the fully 2D simulations (Fig. 2-31), with marginal differences.

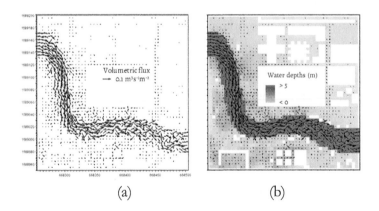

(a)   (b)

Fig. 2-32. Two examples of coupled 1D-2D simulated flood maps: (a) amplitudes and directions of flood-wave propagations; (b) floodwater depths with amplitudes and directions of flood-wave propagations

This coupled 1D-2D model allows a high level of flexibility when using lateral links. It is particularly useful for simulating overflow from a river channel onto a floodplain. Moreover, their simulation results can be represented in both 1D and

2D visualisations (Section 2.2.8 and 2.4.6, resp). In this research, the coupled 1D-2D modelling was chosen as a main urban-flood modelling tool for two case studies: Kuala Lumpur (Chapter 6) and Ayutthaya (Chapter 7).

2.7 Issues concerning complex-urban flood modelling

2.7.1 Complex topography

An appropriate quality of urban-flood modelling results crucially depends on essences of input data that can replicate characteristics of topography and its surroundings as close to reality as possible. Even though the standard top-view topographic data can adequately be used to represent essential urban features, some alleys in between buildings, underpasses, and/or kerbs still cannot properly represent. The geometric discontinuities of such urban features can play a significant role in diverting shallow flows in complex urban environments. The floodwater-flow patterns may result in interactions between these matters, but the way they interact remains mostly unspecified.

Generally, when resolutions of computational-grid schemes in urban flood modelling have increased (a space of each computational grid approaching closely to zero), some research (Marks & Bates, 2000; Horritt & Bates, 2001; Haile & Rientjes, 2005b) show that it can have considerable effects on the simulated results of inundation extents and timing predictions. However, enhancing schematic resolutions may not always achieve better quality of simulated results. When using high-resolution top-view LiDAR data as input, schematics of urban-flood modelling can plausibly be created, yet applying such schematics may have difficulties in representing flows through these complex urban features. In an urban flood simulation, floodwater flows may have been blocked by high trees, alike-connected buildings, and/or overreaching structures (Fig. 2-33a). In reality, buildings are typically built close together or even connected to each other. The actual flood pathways may be hidden underneath such connecting urban features (Fig. 2-33b), e.g. elevated roads, sky-train tracks, bridges, arches, and overarching structures. Razafison et al. (2012) noted that these (missing) structures and pathways could have a major impact on flood propagation and flow dynamics.

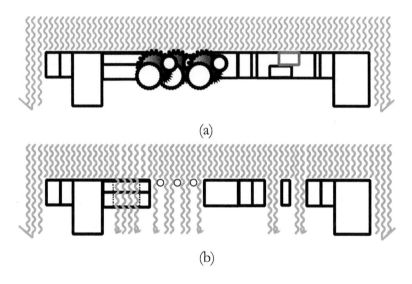

(a)

(b)

Fig. 2-33. Two conceptual maps of (a) predicted inundation when using conventional top-view topographic data as input; (b) actual inundation when floodwaters can freely flow through

The behaviours of flood propagations are dominantly associated with flow routeing processes. Directions of floodwater flows are not only controlled by topographic terrains but also conveyed by urban structures above its terrains. Changes in elevation surface can have a profound impact on processes such as overland flow and human activities (Mitasova et al., 2011).Without doubt, identifying and describing key components of a complex urban area are crucial. Inadequately analysing these complexities may lead to inadequate flood-protection measures (Fig. 2-34) or even lead to more catastrophic situations, especially for such a complex city.

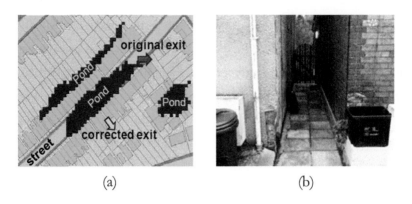

(a)   (b)

Fig. 2-34. (a) An example of a simulated result using conventional top-view LiDAR data
as input for 2D flood model; (b) A small alleyway (source by Boonya-aroonnet, 2010)

From Fig. 2-34a, a red arrow shows a simulated result of floodwater flows toward
the Northeast direction. From Fig. 2-34b, a yellow arrow shows an alley as a true
flood pathway, but none of simulated floodwaters flows toward the Southeast
direction. Due to these true flood pathways were missing and they were not
incorporated in these 2D flood-model schematics. Incorporating these (missing)
pathways in 2D model schematics should be adequate to represent such hidden
openings from surrounding complex urban features. The scale factors of
developing such appropriate details of (fine or coarse) 2D model schematics are
crucially depending on final resolutions of each flood simulation application and
purpose.

2.7.2 Submerge drainage systems

Further descriptions of submerge drainage systems are beyond the scope of this
research and further details can be found elsewhere (e.g. Vojinović et al., 2006;
Vojinović & Van Teeffelen, 2007; DHI, 2016a; DHI, 2016c). Due to many local
drainage systems in most developing cities are commonly not extensive and not
well maintained. Even if an information of designed drainage systems is available,
it may be hard to examine actual existing capacities of such drainages buried

underground and it may need a lot of investment surveying costs and times. However, some basic concept of applying 1D, 1D dual drainage, coupled 1D-1D modelling, and coupled 1D-2D modelling for submerged drainage systems are given hereafter.

2.7.2.1 1D drainage pipes

As mentioned in Section 2.2.5, De Saint Venant flow equations (Cunge et al., 1980) are commonly used to describe changes in flow velocities and water depths to both open channels and drainage pipes, which their capacities are not completely full. However, a pressurised flow can exist in storm pipes, which become surcharged during rainfall (Fig. 2-35).

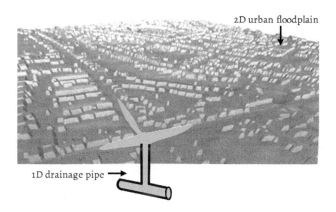

Fig. 2-35. An example conceptual map of a 1D pipe network and a 2D urban floodplain (source by Vojinović & Abbott, 2012)

Modelling pressurised pipes is one of the most difficult aspects. Yet their fundamental problems and no equations have been developed, which adequately represent a transition from the free surface flows to the pressurised flows. Typically, pipes are predominantly free surface and only rarely surcharged (Fig. 2-36). They could be modelled by using the Preissmann (1961) slot approximation.

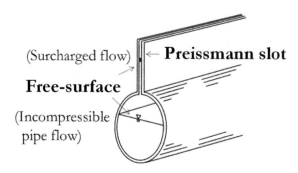

Fig. 2-36. An example of Preissmann (hypothetical) slot (source by Ukon et al., 2008)

This approximation uses a narrow slot, which runs the length of the conduit and extends vertically up to infinity. The slot allows a free surface to be maintained and therefore a transition does not occur. The Preissmann Slot is not recommended in pipes when they are permanently and/or heavily surcharged, i.e. highly pressurised. In this respect, a dedicated pressurised pipe model must be used, which does not include the Preissmann Slot or base flow, approximations. There are two common options for doing this by using pressure or force main solution models, both specified as part of the conduit information. The more information of pressurised pipe models are available elsewhere (Chaudhry, 1979; Reeves, 2013).

For some complex cities, issues of topography definitions for subsurface drainages may often seem to be time-consuming for both the surveying and processing. Due to the fact that these drainage systems were planned on the desk, but these complex systems were built and buried under the ground, which hardly to know the precise locations and actual conditions of them. Typically, simplified pipes and manholes (Fig. 2-37a) are adequately used to replicate common flooding. Whereas flood flow behaviours in the actual pipe definitions could be far more complex (Fig. 2-37b).

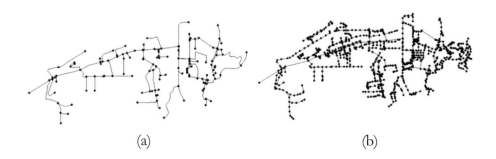

(a) (b)

Fig. 2-37. Examples of the urban drainage system of Coimbra, Portugal, represented in 1D models; (a) a simplified 1D drainage pipe network, (b) an actual 1D drainage pipe network (source by Simoes et al., 2010)

2.7.2.2 1D dual drainage concept

A dual drainage concept could be used for the complex surface and subsurface systems. A connection between two drainage systems is linked through the catch pits (Fig. 2-38b), whereas the minor system is typically connected through the manholes. When floodwater enters to the sewer lines through the catch pits, the pits act as a temporary storage area before the water is finally discharged into the manholes. To allow the inundation to be connected in a proper (physical) way, the major and minor systems must be coupled appropriately; the different small changes in the elevation of the channels and gradient pipe slopes may be accounted for significant gravity-driven flows in the simulation results.

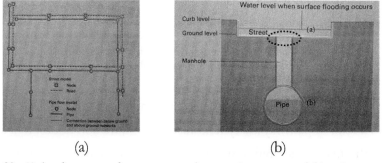

(a) (b)

Fig. 2-38. (a) An aboveground storage network as a major system and (b) a pipe network as a minor system in a 1D dual-drainage modelling setup (source by Vojinović & Abbott, 2012)

2.7.2.3 Coupled 1D-1D modelling

While some urban flooding behaviours are more complicated than others, especially in the conjunction areas between the surface channels and subsurface conduits. The designated surface channels (major system) and subsurface channels (minor system) are synchronised at conjunction nodes. This interaction takes place via the links between flows above and below ground – inlet as a surface link and manhole as a subsurface link.

In the coupled 1D-1D modelling setup, when pathway channels are synchronised with conventional open channels, their flows could be simulated by coupled two or more 1D models together (examples in Fig. 2-39). Each of system is modelled with the sets of the 1D equations, and regularly synchronising water exchanges between them. The exchange flows between the systems could also be modelled by using broad-crested weir equations (Evans et al., 2007).

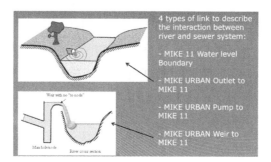

Fig. 2-39. The example of the coupled 1D-1D models: (a) pipe networks coupled with the open channels (b) a weir construction coupled with the open channels (source by DHI, 2016c)

The underground pipe network (minor system) and the designated channel (major system) of a road or pathway are synchronised at conjunction nodes. In a complex city, both the major and minor systems can be simulated in a coupled 1D-1D model (Fig. 2-40).

Fig. 2-40. The complex surface channels (blue colour) and subsurface conduits (black colour) represented in a coupled 1D-1D modelling setup of Coimbra, Portugal (source by Simoes et al., 2010)

The connections for the major system could be linked through the catch pits, where the minor system is connected through the manholes. These models provide fast insight in flow paths and depressions based on a digital elevation model (Simoes et al., 2010; van Dijk et al., 2014). To allow inundations to occur in a proper (physical) way, major and minor systems must be coupled properly; some small changes in elevations of each channel and gradients of pipe slopes may be accounted for great gravity-driven flows in the simulation results. Thus, the best possible setting up all pathways and ponds for 1D-1D models should be carefully made by using maximum information, e.g. DEMs and land use data (Zevenbergen et al., 2010).

2.7.2.4 Coupled 1D-2D modelling with subsurface networks

Applying dual drainage concept of coupled 1D-1D models may be adequately used for some pluvial flood situations when topography definitions of each cross section are well defined. However, such simulated results are limited by their (well defined) cross-sections, yet the floodwater beyond that cross-sections cannot be correctly considered. In this case, the simple 1D subsurface modelling should be coupled to a fully 2D urban flood modelling in order to simulate the exceedance floodwater flows between the pipes and the urban floodplain.

The interaction between the surface and subsurface systems are determined according to the type of link. For example, discharges generated by pumping stations, weirs or orifices are regarded as the lateral inflows to the 2D models. Therefore, if the pipe flow exceeds the ground level, discharges are then computed by the weir (or orifice) discharge equation and it is considered as a lateral inflow in the 2D models.

Even though the complicated subsurface systems are generally buried underneath the complex cities, lacking data in the existing subsurface capacities and the time-series records are still the main issues for modelling setups and calibrations. Therefore, a theoretical study of subsurface systems, which related to the 1D pipe network, coupled 1D-1D, and coupled 1D-2D models, is beyond the scope of this research (see more in Mark et al., 2004).

2.7.3 Control structures for 1D models

Even though 2D models in MIKE 21™ software are capable of simulating flows more than one direction, this software still has limitations of involving hydraulic structures on their 2D models. Owing to this, applying 1D models in MIKE 11™ can enable hydraulic structures to be simulated in their computational nodes. More descriptions of hydraulic structures and compatibility conditions are beyond the scope of this research scope and their further details can be found in DHI (2016a).

When each hydraulic structure is described as an overflow structure, an underflow structure, a radial gate, or a sluice gate, they should be placed on a branch at the Q point. Moreover, pump structures with internal outlet, which increase a local water into the branch, can also place at the Q point. The structure node describes a Q point at the structure (location 'j' in Fig. 2-41) in between two h points at each side (locations 'j-1' and 'j+1' in Fig. 2-41).

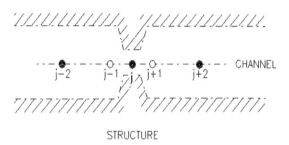

Fig. 2-41. An example of a single branch with a hydraulic structure at 'j' grid or at 'Q' point (black circles), and h points (white circles) at either side of the structure (source by DHI, 2016a)

At each structure node, it could be calculated by replacing a momentum equation with an h-Q-h relationship, an h-Q relationship, or a Q assignment. Unlike numerical descriptions of partial differential equations, no numerical damping of physical processes is introduced, as a compatibility condition that should hold at any point in time. The only error introduced comes from the linearization of the quadratic relation at a new time level. Whereas pumps with external outlet pump water out of the branch placed in h points.

CHAPTER 3
Conventional top-view LiDAR topographic data

Achieving high-resolution topographic data have their own difficulties. Acquiring such high-resolution data using conventional land surveys is typically time-consuming and intensive works. Advanced aerial surveys, in contrast, can provide appropriate details of topographic data with less time-consuming in fields, especially for remote areas. When remote sensing technologies have emerged and then shifted to a digital edge, achieving such high-resolution topographic data with high accuracy is more promising and practical. Amongst of these innovative remote-sensing technologies, an aerial LiDAR system has received much attentions in urban analyses and urban flood-modelling applications. In this chapter, evolution in topographic data acquisition is drawn in Section 3.1. Overviews of top-view LiDAR data acquisition are described in Section 3.2. Analysing and registration processes of raw LiDAR data are explained in Section 3.3. Simplification of top-view LiDAR data is given in Section 3.4. Issues concerning top-view LiDAR data are also discussed in Section 3.5.

3.1 Evolution in topographic data acquisition

Since former times, people had tried to illustrate their surroundings by painting. Some landmarks and some urban-feature shapes in a painting map can easily be recognised, while other objects may hardly be interpreted and sometimes misscaled or misplaced (Fig. 3-1).

Fig. 3-1.   A map of "La Ville de Judia" - Ayutthaya Island, Thailand (source by Struys, 1718)

Nowadays, a modern topographic map can represent both natural and urban features, of which features on the map can easily be seen in reality (e.g. water bodies, roads, railways, cities, and parks). Whereas other features on the map are still hardly detected in reality (e.g. pipelines, tunnels, and administrative boundaries). These modern maps generally created in particular scales. Almost all of them show some legend illustrations and descriptions as in common.

Elevation data (topographies) could be counted as an utmost importance information for a number of applications, e.g. surveying, urban planning, engineering, and flood management. Acquisition topographic data has never been an easy task. Space and time criterions including fundamental aspects of surveying should have been considered and defined carefully to achieve both quality and quantity of topographic data. Such criterions should be determined before approaching to acquisition technology selection, surveying, and data processing steps. When people desired to perceive the Earth "as birds do", they have started to lift a camera off the Earth's surface, trying to capture some aerial scenes. Neubronner (1909) was one of them; he even sought to implement a lightweight camera to (trained) pigeons, aka pigeon photography (examples in Fig. 3-2).

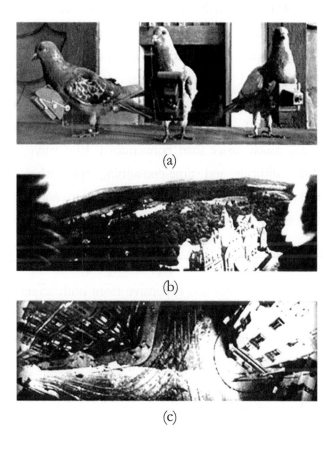

(a)

(b)

(c)

Fig. 3-2.   (a) The trained pigeons were installed with cameras. Some pigeon photos were taken at (b) Schlosshotel Kronberg's oblique-view and (c) Frankfurt's top-view, Germany (source by Neubronner, 1909)

However, two main issues of consistency and reliability in the pigeon photography were too difficult for solving, either during that time or even now.

Typically, a conventional land surveying can appropriately provide an accurate elevation for each accessible location. Since balloons and aeroplanes were invented, they have become more widely used for Earth's observation. In 1858, the formally known of France's aerial photographs were first taken from a balloon by Gaspard-Félix Tournachon (Colwell, 1997). In 1908, some aerial photographs were later taken from an aeroplane by Bonvillain, L. P., piloted by Wright, W. (Jensen, 2000). In the mid-20th century, stereo-photogrammetry techniques were then implemented for topographic reconstructions using stereoscopic equipment. After then, elevation points and contour lines can be reconstructed from these overlapping aerial photos, either the monotone or colour photos can adequately be used (Ham & Curtis, 1960). Even though observing the Earth's surfaces by an aerial surveying are much faster than conventional land surveys, a film-based approach was still difficult and time-consuming approaches for both the photography and stereo photogrammetry processes. Registering those overlapping aerial photos necessarily yet required many ground control points (GCP) for orthophotography process in order to create an analogue topographic map (Zahorcak, 2007). During that machine-age era, adopting such conventional stereo-photogrammetry techniques was too expensive from both acquisition tools and mapping equipment. It may also be extremely laborious and time-consuming approaches to achieve high-resolution topographic map by adopting such analogue-based stereo-photogrammetry techniques.

In the late-20th century, most analogue topographic formats have been shifted into digital file formats. Since then, advances digital topographic data have been dramatically improved, processing, re-processing, transferring, and storing such digitally topographic data can be handled much easier. In the early-21st century, several topographic data can be digitally obtained by using new remote sensing technologies, e.g. radar, interferometry SAR, and LiDAR. Amongst of these sophisticated technologies, an aerial light detection and ranging (aerial LiDAR) or airborne laser scanning (ALS) system can be counted as the most popularly use for urban analyses. Nevertheless, the first LiDAR applications originated from meteorology fields, when the National Centre for Atmospheric Research applied them to measure distances and thicknesses of the clouds (Goyer & Watson, 1963). Eventually, accuracies and usefulness of a LiDAR system for topographic observations were acknowledged, when they were implemented to map the Moon's surfaces in the Apollo 15 mission in 1971 (Rodionov et al., 1971).

3.2 Top-view LiDAR data acquisition

A new era of remote sensing technologies allows surveyors to obtain topographic data from a remote distance without touching target objects. Many improvements in remote sensing technologies are now capable of providing digital topographic data in high resolutions and accuracy. Most remote-sensed topographic data are commonly obtained from top viewpoints by either airborne or even spaceborne platforms. Amongst of these top-view topographic data acquisitions, an aerial LiDAR system has long received considerable attentions due to their reliabilities and precise measurement capabilities. In the late 1970s, when acquisition top-view LiDAR topographic data had started, the accuracy of such top-view topographic data was still limited due to inadequate determinations of the aircraft positions (Ackermann, 1979). Since the late 1980s, when advances in global positioning systems (GPS), inertial navigation units (IMU), laser scanning systems, and microcomputers have dramatically been improved, obtaining top-view topographic data by adopting aerial LiDAR system can be promising to achieve high resolutions, better accuracies, and more reliabilities (Gordon & Charles, 2008). As a result, during the past decades, an extensive used of top-view LiDAR data have rapidly expanded in a field of urban spatial analysis (Haala et al., 1998), e.g.

terrain analysis, urban planning, architecture, and using as a topographic input data for urban flood modelling. In this research, top-view LiDAR acquisition is also given (Fig. 3-3)

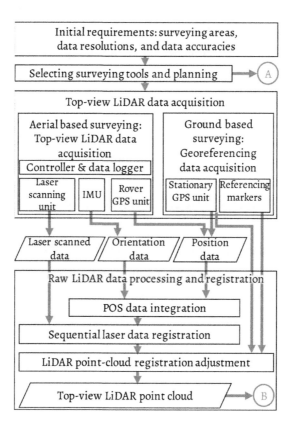

Fig. 3-3. A workflow of top-view LiDAR data acquisition ('B' and 'A' descriptions in Section 3.4 and 4.2, resp)

3.2.1 Aerial based surveying for the top-view LiDAR data acquisition

When applying an aerial LiDAR system as an acquisition tool, a special or modified aircraft is commonly used to carry all heavy and sophisticated surveying tools, e.g. a laser scanning unit, a rover global positioning system (GPS) unit, an inertial measurement unit (IMU), and a controller with data logger unit (Fig. 3-4). In the aircraft, all these tools need to be precisely installed and mounted on rigid platforms with stabiliser rigs to minimise geometric errors of equipment movements and instabilities.

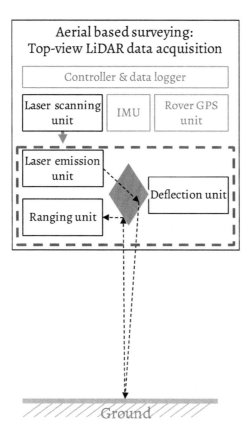

Fig. 3-4.   The mandatory equipment of laser scanning unit in an aerial LiDAR system

In a laser-scanning unit, while a laser unit is emitting laser beams, a deflection unit distributes these beams to ground surfaces covering a large swath of a flight track. The laser beams scattered on the ground can characterise the laser patterns from this deflection unit. These beam divergences, therefore, show a unique pattern in a final raw laser-scanned data dependently varying for each detection type (see examples in Wehr, 2008). For example, the two case studies in this research used different aerial LiDAR system: RIEGL® LMS-Q560™ for Kuala Lumpur (Section 6.2) and LEICA GEOSYSTEM™ ALS70™ for Ayutthaya (Section 7.2). While the ALS70™ aerial LiDAR system uses oscillating mirror for the deflection unit, the matter of the scan patterns can be set as parallel, triangle, or sinusoid (Fig. 3-5a). Whereas the Q560™ system uses a continuously rotating optical polygon, therefore, its scanned patterns can be represented as a rotating elliptical over the ground (Fig. 3-5b).

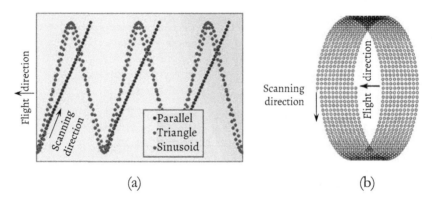

(a)   (b)

Fig. 3-5.   Two examples of different deflection units and its scanned patterns:
(a) parallel, triangle, sinusoid in ALS70™, and (b) elliptical in Q560™

When the laser scanner receives backscattered signals from the surfaces' targets, a ranging unit determines a distance from each returned signal. Generally, two of the most commonly used in ranging units are time-of-flight and phase-shift systems. A time-of-flight system determined a range of the aircraft to the target by calculating both travel times and shooting laser-beam angles of each returned laser signal. Whereas a phase-shift system uses a phase difference between emitted and backscattered signals to determine the distance. In theoretical, the phase-shift system is more accurate. However, the time-of-flight system has longer range capabilities than the phase-shift system (Puente et al., 2013). Nowadays, various types of the aerial LiDAR system in the market (see examples in Maune, 2007) can simultaneously measure distances and instantly store the raw laser-scanned data into a data logger device (Fig. 3-6).

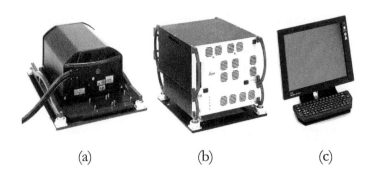

(a)   (b)   (c)

Fig. 3-6. Aerial-based LiDAR instruments of Leica Geosystems: (a) Leica ALS70 laser scanner, (b) controller and data logger, (c) operator interface with keyboard

During an operation of aerial surveying, a pilot has to maintain a most-likely constant flight height and side-lap conditions in between 30% (Fig. 3-7) to 60% of each overlapping flight path to achieve a (constant) area of acquisition strip on the ground.

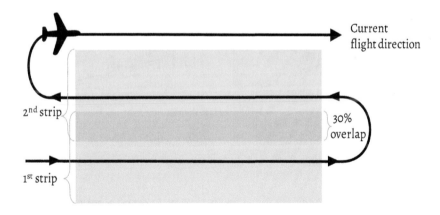

Fig. 3-7. An example of 30% side lap between two strips of LiDAR data

A width of the acquisition strip on the ground can vary from 200 m to 2 km. The range of laser pulses can be a long distance from several metres up to kilometres. The rate of data capture scans can be 500 points s^{-1} or more points of scanned lines per second. These conditions can characterise quality and resolution of top-view

LiDAR data. A visible area related to quality of top-view LiDAR topographic data is crucially depending on speeds, flight heights, and angle of scanned receptions. As a result, the details of top-view LiDAR point cloud may vary in the different point cloud densities (Fig. 3-8).

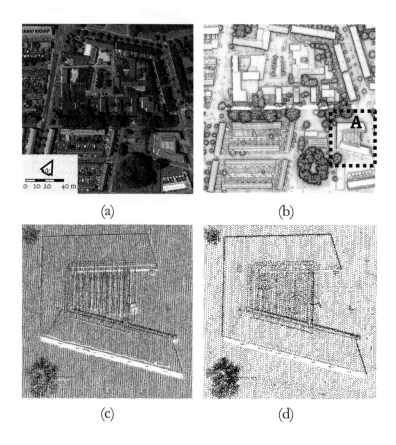

Fig. 3-8.   An example of (b) top-view LiDAR point cloud (AHN 2, 2014) obtained in the complex urban scene (Google Earth™ 7.1.5.1557, 2016a). Two examples of different LiDAR point cloud densities: (c) ~4 points per m² and (d) ~2 points per m² (location 'A' in Fig. 3-8b)

3.2.2 Aerial based surveying related to the ground

Besides the aerial based surveying (Section 3.2.1), the ground-based surveying is also imperative in top-view LiDAR data acquisition. While a rover GPS unit, IMU, and laser scanning system are operating (onboard) in the aircraft, the known georeferencing locations are simultaneously recorded on the ground using stationary GPS unit (Fig. 3-9).

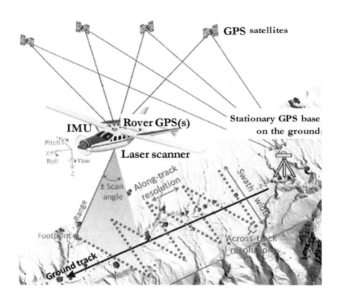

Fig. 3-9.   A special aircraft is commonly required to carry the heavy load of LiDAR, GPS, and IMU instruments. The schematization shows an example of a zigzag sinusoidal pattern distributed from an aerial laser-scanning unit

When the rover GPS unit and the stationary GPS unit are simultaneously operated, both the rover and stationary GPS unit (Fig. 3-10) continuously receive and record the signals from the same set of GPS satellites at all times. For each recorded time, the differential calculations of the rover and stationary GPS signals can precisely estimate the georeferenced position of the aircraft. Typically, an aircraft flight path must be in a range of 25 km maximum for each of the stationary GPS unit. In a large study area, the multiple stationary GPS units can also be applied. These multiple stationary GPS units should be adequately distributed to maintain the maximum ranges for all flight paths.

Fig. 3-10. An example of GPS stationary survey (source by Johnson, 2004)

3.3 Raw LiDAR data processing and registration

When the post processing of differential GPS (DGPS) approach is performed, the differential GPS signals obtained from both the rover and stationary GPS units can be used to calculate the sub-metre aircraft positions. In the meantime, the IMU also records orientations of the aircraft. When GPS signals are weak, the recorded orientations from the IMU can be used to substitute the missing positions. Both the position and orientation records are also known as the data of position and orientation system (POS). In the in post-processing steps, the raw laser-scanned data can sequentially register with the POS data and the georeferenced topographic elevations can then be created. After that, a strip of a top-view LiDAR data can be created. Sometimes, both the laser-scanned and POS data have their independent units. They can be generated by different products with different software. Therefore, the combinations of these data may vary and they should be carefully concerned (Wehr & Lohr, 1999).

In the consecutive LiDAR strips, when the correspondences between of two or more strips were required, they must be carefully defined in the same georeferencing system, e.g. world geodetic system 1984 (EPSG 4326: WGS 84). In this respect, the ground control points (GCPs) obtained from the stationary GPS station. By using these GCPs as geo-referencing locations, the overlapping LiDAR strips can then be re-adjusted precisely in the same georeferencing location. Therefore, the registered composition of different strips can finally be matched in the same scale, alignment, and geolocation (Bellocchio et al., 2013). To check the differences between a pair of survey registered LiDAR data, another set of GCPs can be performed by differential percentile of a cross-verification (%Diff).

$$%Diff = \frac{Xo - Xs}{Xo * 100} \qquad \text{Eq. 3-1}$$

where Xo is observed value, and Xs is the surveyed (or simulated) value. To check effectiveness of the registered LiDAR data, another set of GCPs can be performed. The residuals from the cross-verification procedure were used to calculate following four verification measures (Knotters & Bierkens, 2001): (i) the systematic error or mean (ME):

$$ME = \bar{x} = \frac{1}{n} \sum_{i=1}^{n} x_i \qquad \text{Eq. 3-2}$$

where n is the number of observed value, and x_i is the differences between observed to surveyed (or simulated) value at the same location; (ii) coefficient of determination (R^2):

$$R^2 = 1 - \frac{\sum_{i=1}^{n} (o_i - \bar{x})^2}{\sum_{i=1}^{n} (s_i - \bar{x})^2} \qquad \text{Eq. 3-3}$$

which is the coefficient of determination of the surveyed values, where o_i is observed values s_i is surveyed value; (iii) the root mean squared error (RMSE):

$$RMSE = \sqrt{\frac{1}{n}\sum_{i=1}^{n}(x_i)^2}$$

Eq. 3-4

which is a measure of the overall closeness of surveyed (or simulated) to observed elevations; and (iv) the mean absolute error (MAE):

$$MAE = \frac{1}{n}\sum_{i=1}^{n}|x_i|$$

Eq. 3-5

which is also a measure of accuracy, but is less sensitive to large errors than RMSE (Yuan et al., 2008). Both MAE and RMSE can be used for variation diagnoses in the errors for a set of forecasts; lower values are the better accuracy. The RMSE will always be equal or larger to the MAE; the greater difference between them, the greater the variance will be in the individual errors in the sample. If RMSE is nearly the same as MAE, then all the errors are of the same magnitude. However, only the last three verification measures, i.e. ME, R^2, and RMSE, were used for the rest of this research.

3.4 Top-view LiDAR data simplification

From Section 3.2 (location 'B' in Fig. 3-3), a registered top-view LiDAR point cloud contains essential topographic information, which can be used in ranges of applications. Nowadays, processing such digitally topographic data can be handled much easier. However, an extremely large number of points need to be analysed and simplified carefully, while key components of urban feature characteristics essentially need to be maintained. To simplify and achieve a manageable top-view LiDAR point cloud, the two simple steps, i.e. a key component extraction and a point cloud rasterization (Fig. 3-11) are described hereafter.

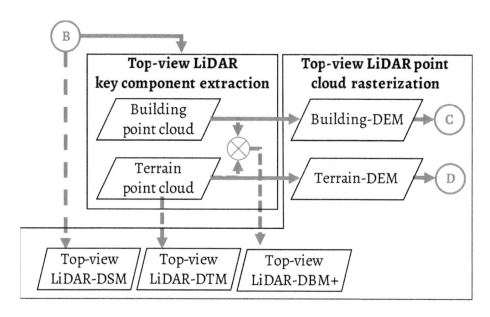

Fig. 3-11. A workflow of the data simplification using top-view LiDAR point cloud (location 'B' referred from Fig. 3-3). The further steps of terrain and building point cloud processing are given (descriptions 'C' and 'D' in Section 5.3)

3.4.1 Top-view LiDAR point cloud extraction

Extracting key components of the raw top-view LiDAR point cloud data has been widely discussed in different fields and applications. During the last decades, many filtering algorithms have been explored and developed for classifying top-view LiDAR point cloud in order to extract some key components of urban features, e.g. land covers (Yan et al., 2015), trees (Alonzo et al., 2014; Han et al., 2014; Chen et al., 2015), buildings (Kabolizade et al., 2010; Awrangjeb et al., 2013; Mongus et al., 2014; Song et al., 2015; Ferraz et al., 2016), roads (Li et al., 2015; Ferraz et al., 2016), or even vehicles (Yao et al., 2010). When a set of criteria has been characterised, essential information embedded in point cloud can be extracted and classified into particular segments.

Some recent research applied the different filtering algorithms, e.g. morphological filtering and adaptive triangulated irregular network (TIN) filtering, to extract terrains from top-view LiDAR point cloud. In morphological filtering, when adjacent ranges of referencing elevations are given, points in these elevation thresholds can be classified as a terrain point cloud, whereas other points beyond these thresholds can be classified as a non-ground point cloud.

Even though this morphological filtering is simple and straightforward, determining correct elevation thresholds is still difficult, especially for hilly areas. Too high a threshold may lead some trees and small buildings to be incorporated into the extracted terrain point cloud. Too small a window size may also lead some flatted rooftops to be classed as terrains (Kilian et al., 1996). In adaptive TIN filtering, while neighbouring points assume to have a similar value, the more distant points, in contrast, assume to have the more different values. When the surface distances of TIN are defined, points of which that meet these triangle criteria can be added, during each iteration. The iterative processes end when there are no points that matched the threshold range. This filtering can adequately remove small buildings and bridges. However, some weakness in removing big buildings are still difficult. Moreover, some filtering methods consume a lot of computational time (Elmqvist et al.; Axelsson, 2000; Zhou et al., 2004).

In this research, a straightforward approach can be adequate to extract and classify key component of terrains from surrounding urban features. When the study area is relatively flat and small, a referencing elevation and a maximum-minimum threshold can be simply determined. When a set of such criterions have been defined, the elevation points in these thresholds can then be extracted and classified essential terrains from top-view LiDAR point cloud. A conceptual algorithm of terrain point cloud extraction is presented, as follow (ALG. 3-1).

ALG. 3-1. A conceptual algorithm for top-view LiDAR terrain extraction

DATA Raw top-view LiDAR points, rawLiDAR
BEGIN
    SET Initial terrain elevation, iniTerrain //e.g. 4 m msl
    SET Terrain threshold maximum, thresholdMax //e.g. +1 m referred to iniTerrain
    SET Terrain threshold minimum, thresholdMin //e.g. -2 m referred to iniTerrain
    SET Selected LiDAR-terrain points, terrainLiDAR = ""
    SET Un-selected LiDAR points, unselectedLiDAR = ""
    SET Key feature field to all points, keyFeature = ""
    SET Projection plain, projPlain = z
    PROJECT rawLiDAR-points along projPlain
    WHILE projPlain = z;
        IF rawLiDAR-points in (iniTerrain AND
        thresholdMax AND thresholdMin)
            KEEP rawLiDAR-points, terrainLiDAR
            ASSIGN keyFeature = "terrain", terrainLiDAR
        ELSE
            KEEP rawLiDAR-points, unselectedLiDAR
        END IF
    END WHILE
END

When the study area is relatively flat and small, essences of terrain point cloud (Fig. 3-12b) can be extracted from the raw top-view LiDAR data (Fig. 3-12a).

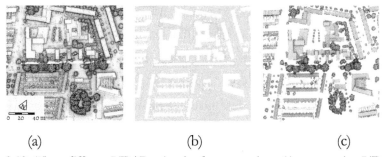

(a)   (b)   (c)

Fig. 3-12. Three different LiDAR point cloud segmentations: (a) raw top-view LiDAR point cloud, (b) terrain LiDAR point cloud, and (c) mixed building and tree features of non-ground point cloud

Whereas the rest of remaining point cloud can be segmented into a non-ground point cloud, e.g. buildings and trees (Fig. 3-12c). For the building extraction, the distinction of such mixed building and tree point cloud could be one of the most challenging task. Due to top canopy of high trees may cover some parts or even the whole of building rooftops. Typically, the returned pulses of top-view LiDAR data may contain with densely random points distributed over rooftop surfaces. Some recent research found that an appropriate combination of images and LiDAR data could improve performances of building rooftop (or building footprint) extractions (Schenk & Csathó, 2002; Habib et al., 2008). Yet, in image-based building extraction, some remaining regions near the building edges may produce inaccurate building footprint boundaries. These errors can be attributed either to radiometric artefacts introduced during the orthorectification process (resampling) or to building shadows (Sohn & Dowman, 2007). Even though extracting edge outlines of such rooftops could be easy when trees not cover the rooftops, it may still be difficult to exact the rooftop boundaries when such rooftops are covered by trees (Zhang et al., 2006). Moreover, some visual interpretation approaches using high-resolution image satellites may often use to analyse the more complex building-footprint boundaries.

Although these approaches are labour intensive (Gross & Thoennessen, 2006), it is practical for some small to moderate sizes of the complex city. In this research, the mixed building and tree features in non-ground point cloud (unselectedLiDAR referred from ALG. 3-1) can be extracted by using the given building footprint boundaries in GIS formats. Such building-footprint boundaries were generated by visual interpretation approaches. A conceptual algorithm of terrain point cloud extraction is presented (ALG. 3-2). This simple algorithm can be chosen when the building footprint boundaries are given. However, reconfirmation of these boundaries using the latest update remote-sensing satellites analysing or even ground truth surveying is crucial.

ALG. 3-2. A conceptual algorithm for top-view LiDAR building extraction

DATA Un-selected LiDAR points, unselectedLiDAR
BEGIN
    SET Boundary of building footprint, buildingBound
    SET Selected LiDAR-building points, buildingLiDAR = ""
    SET Selected LiDAR-tree points, treeLiDAR = ""
    SET Key feature field to all points, keyFeature = ""
    SET Projection plain, projPlain = z
    PROJECT unselectedLiDAR-points along projPlain
    WHILE projPlain = z;
        IF unselectedLiDAR-points in buildingBound
            KEEP unselectedLiDAR-points, buildingLiDAR
            ASSIGN keyFeature = "building", buildingLiDAR
        ELSE
            KEEP unselectedLiDAR-points, treeLiDAR
            ASSIGN keyFeature = "tree", treeLiDAR
        END IF
    END WHILE
END

When using such given boundaries, the building point cloud (Fig. 3-13b) can be extracted from the non-ground point cloud (Fig. 3-13a). Whereas the rest of point cloud can be segmented into mixed trees point cloud (Fig. 3-13c).

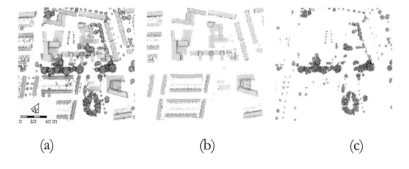

(a)   (b)   (c)

Fig. 3-13. Three different LiDAR point cloud segmentations: (a) the remained point cloud, (b) building LiDAR point cloud, and (c) mixed tree point cloud

3.4.2 Top-view LiDAR rasterization

Even though some top-view LiDAR point cloud data are available, most topographic data nowadays are still provided in the regular square-matrix structured grids. Such raw or extracted top-view point cloud (Fig. 3-14a) can be commonly simplified and transformed into the square-grid topographic data (Fig. 3-14b), known as a rasterization process. The rasterization is also handy for further analysing, processing, manipulating, and delivering the data (Martz & Garbrecht, 1992). Due to the square structure grid is straightforward and it usually used less time for data preparations and processing (Fairfield & Leymarie, 1991).

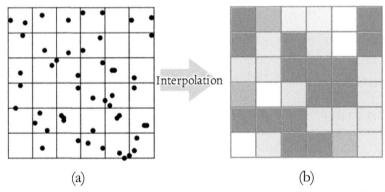

Fig. 3-14. Two conceptual maps of (a) elevation points allocated in a structured grid and (b) interpolated elevation points of square-grid topographic data

Typically, two-dimensional matrix arrays contain an interpolated value (e.g. the elevation) in each of square grid cell. The geographical locations are also implicit and embedded in row and column arrays when the boundary coordinates of these arrays and the regular cell sizes are known. Some geometrical interpolation algorithms can be commonly applied for most up- and down-scaling processes to decide which candidate points will be situated into new grid cell or not. Estimating these unknown intermediate values can be achieved by interpolating the neighbouring known values (blue cells in Fig. 3-14b), and empty cells for non-neighbouring values (white cells in Fig. 3-14b).

For determining these unknown values, a simple linear interpolation can be used for a homogeneous area, due to their this linear interpolation is fast, but their results may have fewer precision (Axelsson, 1999). It must be wise to use the more complex interpolation algorithms for more complex geometry. Inverse distance weighting (IDW), Kriging, and Spline algorithms are some of common interpolation algorithms. When topographic point cloud data are sparse, such algorithms could become more sensitive to replicate spatial cells of local surfaces. They may have some difficulties for the local grid refinements (Joe & Nigel, 1993; Filippova & Hänel, 1998; Durbin & Iaccarino, 2002). When topographic point cloud data are dense, applying Spline appears to show curve and smooth surface output cells. While using sophisticate-weighted average in Spline algorithm could be appropriately used for explaining variations in surfaces, their generated spatial cells can still be exceeded value range of samples and these cells may not pass through such samples. Applying IDW could be adequately used to represent proper details of local surface variations. The IDW algorithm is simple and relatively easy to explain. This algorithm determines an output cell value using a linear weighted combination set of samples. Such assigned weights can be applied as function of distances from an output cell location relative to input points. However, greater distances can result in less influences to output cell values. Other interpolation algorithms can be found elsewhere (Price & Vojinović, 2011). In this section, the extracted top-view LiDAR point cloud were used to create three different types of digital elevation models (DEMs): (i) digital surface model (DSM), (ii) digital terrain model (DTM), and (iii) digital building model with DTM (DBM+), employing the IDW algorithms in the rasterization process. While the DSM represent all top surface elevations, e.g. top of canopies, the rooftop of buildings, the DTM, in contrast, represent only the bare Earth's surfaces of the ground, aka the terrain elevations. A distinguish between DSM and DBM+ is that the DSM typically corresponds all returned pulses of the raw top-view LiDAR data without filtering processes (Rottensteiner & Briese, 2002). The DBM+ corresponds to both the terrain and building key components (Fig. 3-15).

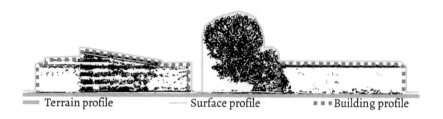

Fig. 3-15. A longitudinal profile of top-view LiDAR point cloud represented in
three different key components: terrains, surfaces, and buildings

In this research, the all returned pulses of the raw top-view LiDAR data are used
to create the raster-based topographic DSMs at different resolutions. A conceptual
algorithm for top-view LiDAR-DSM rasterization is presented (ALG. 3-3).

ALG. 3-3. A conceptual algorithm for top-view LiDAR-DSM rasterization

```
DATA Raw top-view LiDAR points, rawLiDAR
BEGIN
        SET Raster grid size, sizeGrid //e.g. 1 m, 5 m, 10 m, or 20 m grid resolutions
        SET Sampling criteria, samplingPoint = maximum
        SET Sampled point value, sampledValue = ""
        SET Projection plain, projPlain = z
        PROJECT rawLiDAR-points along projPlain
        WHILE projPlain = z;
                COUNT (rows, columns) along projPlain
                FOR each (row, column) do
                        SAMPLING rawLiDAR-points
                        with samplingPoint-criteria, sampledValue
                        CREATE-DEM by interpolating sampledValue with
                        sizeGrid-criteria, LiDAR-DSM
                END FOR
        END WHILE
END
```

The random points of the raw top-view LiDAR data can often be delineated into the simplified raster-based topographic DSMs. These raster-based topographic data contain the matrix arrays with the interpolated value in each cell (Fig. 3-16).

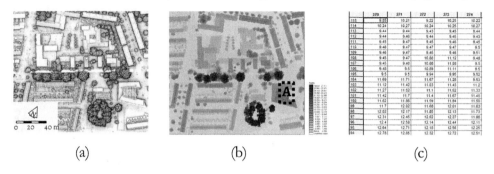

	270	271	272	273	274
115	9.35	10.21	9.22	10.25	10.22
114	10.24	10.27	10.24	10.25	10.27
113	9.44	9.44	9.43	9.45	9.44
112	9.44	9.45	9.44	9.45	9.43
111	9.45	9.47	9.45	9.46	8.45
110	9.46	9.47	9.47	9.47	9.5
109	9.46	9.47	9.45	9.46	9.51
108	9.45	9.47	10.86	11.12	9.48
107	9.46	9.48	10.86	11.00	9.5
106	9.48	9.5	10.89	11.11	9.5
105	9.5	9.5	9.94	9.95	9.52
104	11.69	11.71	11.67	11.28	9.53
103	11.12	11.42	11.03	11.43	11.2
102	11.27	11.52	11.1	11.52	11.33
101	11.42	11.7	11.4	11.67	11.45
100	11.42	11.86	11.59	11.84	11.59
99	11.7	12.02	11.68	12.01	11.63
98	12.02	12.17	11.85	12.13	11.73
97	12.31	12.45	12.02	12.27	11.98
96	12.4	12.58	12.14	12.44	12.11
95	12.64	12.71	12.10	12.56	12.25
94	12.76	12.85	12.32	12.72	12.51

Fig. 3-16. (a) A raw LiDAR point cloud, (b) a raster-based top-view LiDAR-DSM, and (c) matrix arrays contain interpolated values (location 'A' from Fig. 3-16b)

In general, the value descriptions of the raster-based topography are matched with the schematic of the 2D computational grid cells, as mentioned in Section 2.4.1. It can be handy to represent all data values in matrix arrays (Fig. 3-16c and Fig. 3-17a), but it should be much better to illustrate all elevation values in perspective view representation (Fig. 3-17b).

Data values are positioned at specific X,Y coordinates

Grid display represents each cell at the level of the mean value

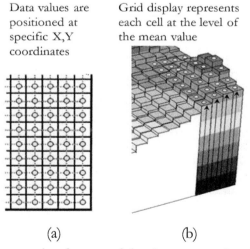

(a)   (b)

Fig. 3-17. Two examples of conceptual elevation representations: (a) a plan view of matrix grid cells; (b) a perspective view of matrix (block) grid cells (source by Berry, 2007)

In the raster-based LiDAR-DSM, higher resolution results will enable the better description of a physical characteristic of the complex cities. Such coarse raster grids may be misrepresented for some key components of the complex urban features and their local impacts may not represent correctly. In contrast, such local key components can be better replicated in high-resolution raster-based topographic data. When a sufficient density of the raw or extracted point cloud, e.g. top-view LiDAR data are available, such estimating unknown values can represent the better details of the local surfaces. For example, the details of urban feature descriptions can represent much better in high resolutions (Fig. 3-18a and b). When the grid sizes are becoming coarsened, there detail descriptions can be smeared or degraded (Fig. 3-18 and d). Therefore, the coarse resolutions may not adequately represent the local characteristic of the urban areas.

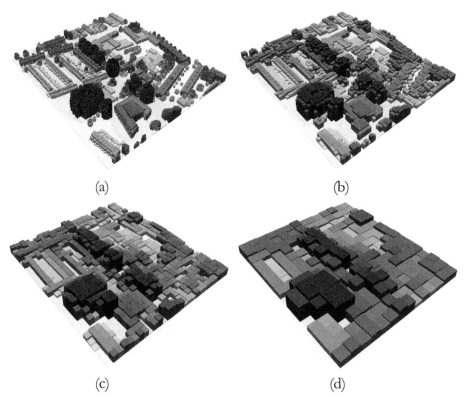

(a)　　　　　　　　　　(b)

(c)　　　　　　　　　　(d)

Fig. 3-18. Four perspective-view representations of top-view LiDAR-DSMs in different grid resolutions: (a) 1 m; (b) 5 m; (c) 10 m; and (d) 20 m grid resolutions

Moreover, when the maximum of elevation corresponding to all returned pulses of the raw LiDAR data was applied, the created DSM certainly represents all top surface elevations without concerning complexity underneath, which can play a significant role in the urban analysis.

When the study area has less complex features, only (flatted) terrain can be used to create a raster-based topographic data. Neither building nor tree point cloud of top-view LiDAR data is incorporated in this flatted terrain point cloud. The extracted terrain point cloud of LiDAR data could be adequately used to create raster-based topographic DTM at different resolutions. A conceptual algorithm for top-view LiDAR-DTM rasterization is presented (ALG. 3-4).

ALG. 3-4. A conceptual algorithm for top-view LiDAR-DTM rasterization

```
DATA Extracted top-view LiDAR points, terrainLiDAR
BEGIN
        SET Raster grid size, sizeGrid //e.g. 1 m, 5 m, 10 m, or 20 m grid resolutions
        SET Sampling criteria, samplingPoint = maximum
        SET Sampled point value, sampledValue = ""
        SET Projection plain, projPlain = z
        PROJECT terrainLiDAR-points along projPlain
        WHILE projPlain = z;
                COUNT (rows, columns) along projPlain
                FOR each (row, column) do
                        SAMPLING terrainLiDAR-points
                        with samplingPoint-criteria, sampledValue
                        CREATE-DEM by interpolating sampledValue with
                        sizeGrid-criteria, LiDAR-DTM
                END FOR
        END WHILE
END
```

While top-view LiDAR-DSM tends to represent many obstructions like top canopies of high trees, top-view LiDAR-DTM (examples in Fig. 3-19b and c) seem to replicate only the bare Earth's surfaces of the ground. Even though some research adopted the LiDAR-DTMs as input for flood modelling (Gallegos et al., 2009), it should be carefully considered when utilising these flatted DTMs as input; the simulated results may not correctly represent floodwater flows and depths (Meesuk et al., 2014). A combination of the building and terrain point cloud data (orange and grey colours, resp. in Fig. 3-20a) can appropriately represent urban structures of the city. In this respect, high trees have been fully removed and not incorporated into this LiDAR digital building model with terrain model (LiDAR-DBM+; see examples in Fig. 3-20b and c). A conceptual algorithm for top-view LiDAR-DBM+ rasterization is presented (ALG. 3-5).

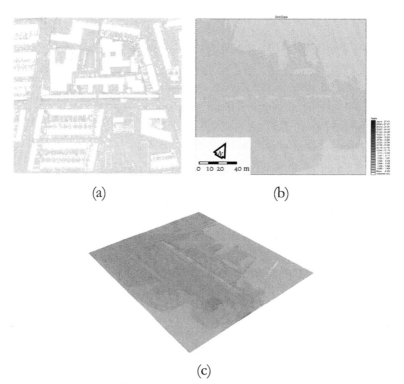

(a)   (b)

(c)

Fig. 3-19. An example of (a) the extracted terrain point cloud of top-view LiDAR data. Two examples of 1 m grid of top-view LiDAR-DTM in different visualisations: (b) a plan view; (c) a perspective view

ALG. 3-5. A conceptual algorithm for top-view LiDAR-DBM+ rasterization

DATA Extracted top-view LiDAR points, terrainLiDAR, buildingLiDAR
BEGIN

    SET Raster grid size, sizeGrid //e.g. 1 m, 5 m, 10 m, or 20 m grid resolutions

    SET Sampling criteria, samplingPoint = maximum

    SET Sampled point value, sampledValue = ""

    SET Assigned DEM-building grid, buildingGrid = ""

    SET Assigned DEM-terrain grid, terrainGrid = ""

    SET Assigned DEM-null grid, nullGrid = ""

    SET Projection plain, projPlain = z

    PROJECT (terrainLiDAR-points AND buildingLiDAR-points)
    along projPlain

    WHILE projPlain = z;

        COUNT (rows, columns) along projPlain

        FOR each (row, column) do

            IF active cell at (row, column) contain

            keyFeature = "building" as majority

                SAMPLING (buildingLiDAR-points) with
                samplingPoint-criteria, sampledValue
                ASSIGN-GRID by interpolating sampledValue with
                sizeGrid-criteria, buildingGrid

                ...

            END IF

        END FOR

        FOR each (row, column) do

            CREATE-DEM use (buildingGrid, nullGrid) with
            sizeGrid-criteria, Building-DEM
            CREATE-DEM use (terrainGrid, nullGrid) with
            sizeGrid-criteria, Terrain-DEM
            CREATE-DEM use (buildingGrid, terrainGrid, nullGrid) with
            sizeGrid-criteria, LiDAR-DBM+

        END FOR

    END WHILE

END

Concerning complexities of urban structure in a city is crucial. However, top canopies of high trees may be perceived as obstructions in urban flood modelling. In this research, the extracted building and terrain point clouds of top-view LiDAR data are used to create the raster-based topographic DBMs+ at different resolutions.

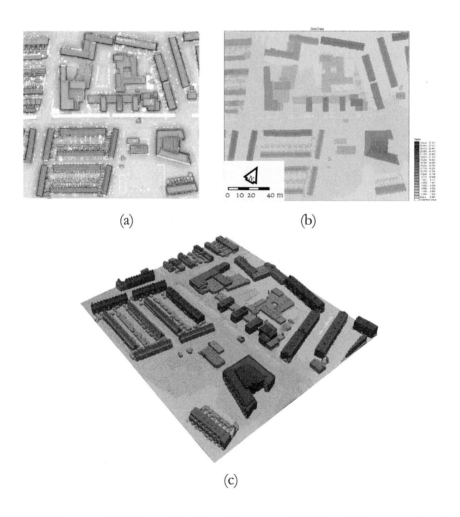

(a)

(b)

(c)

Fig. 3-20. An example of (a) a combination of the extracted building and the extracted terrain point cloud of top-view LiDAR data. Two examples of 1 m grid of top-view LiDAR-DBM+ in different visualisations: (b) a plan view; (c) a perspective view

The final resolutions of raster-based LiDAR-DBM+ should be adequately represented and translated all essences of key features. For further terrain analysis (Section 5.3), other two LiDAR DEMs were distinctively created. The Building-DEM were created from the extracted building point clouds (Fig. 3-21) and the Terrain-DEM from the extracted terrain point clouds (Fig. 3-22).

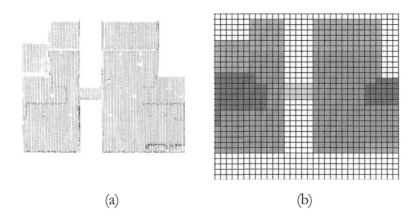

(a)   (b)

Fig. 3-21. Two representation examples of extracted building features: (a) a building point cloud; (b) a raster-based Building-DEM at 1 m grid resolution

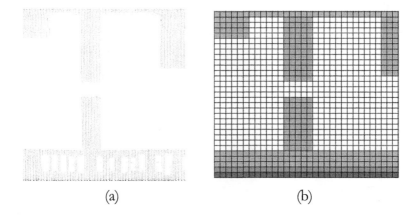

(a)   (b)

Fig. 3-22. Two representation examples of extracted terrain features: (a) a terrain point cloud; (b) a raster-based Terrain-DEM at 1 m grid resolution

3.5 Issues concerning top-view LiDAR data

In top-view LiDAR data acquisition, errors can be caused by the surveying equipment also controllers and data loggers. Such tools can play an important role in achieving a whole set of observed data. For top-view LiDAR topographic data, errors can be caused by aerial based equipment (e.g. laser-scanning unit, IMU, rover GPS units) and ground based equipment (e.g. stationary GPS units and their controllers and data loggers). While errors in surveying equipment could lead to poor quality of observed data, errors in controllers and loggers could lead to losing some observed data or even totally lost the whole data set.

Even though aerial LiDAR data as standard top-view topographic data can adequately be used to represent most features in a city, these standard top-view LiDAR data still have difficulties in representing some urban features, especially for a complex city. In general, top-view LiDAR data are created in high resolutions and obtained by the aerial-based surveying. This aerial LiDAR system pulses the laser signals and scans the target surfaces, e.g. building rooftops and ground terrains, slightly from top views (Fig. 3-23a).

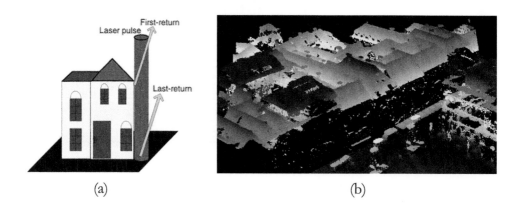

(a) (b)

Fig. 3-23. Two examples of (a) a conceptual graphics of returned laser signals and
(b) a perspective view of missing laser-scanned areas of
building façades in top-view LiDAR point cloud

Therefore, the surfaces of rooftops and/or terrains, which directly exposed to the sky can be scanned easily. Whereas some vertical structures, e.g. alleys, walls, and openings, or some small urban features, e.g. pathways and kerbs, underneath the overarching structures are hard to be detected and still missing in this standard top-view LiDAR data (Fig. 3-23b).

Moreover, the standard top-view LiDAR data are publicly available for some cities, states (e.g. Delaware, Indiana, Iowa, Louisiana, Minnesota, North Carolina, Ohio, and Pennsylvania of USA), or even the whole country's territory (e.g. the Netherlands, Denmark, Finland, and Slovenia), such high-resolution top-view LiDAR data are still costly even in developed countries. The cost of acquiring new top-view LiDAR data for an area of 50,000 ha at a suitable laser pulse density of 6 to 8 points m^{-2} was about three USD per ha plus USD 10,000 to 20,000 for mobilisation costs. This is much higher compared to the cost of acquiring digital aerial imagery for the same area at just over one USD per ha but almost similar mobilisation costs. For example, a case study of Ahtanum, Washington, USA covered 6,689 ha, ignoring mobilisation costs, the LiDAR would cost about USD 20,067 and the imagery would cost about 8,026 USD (Erdody & Moskal, 2010).

CHAPTER 4
Introducing new side-view SfM topographic data

Topographic data can be observed from not only top views but also different viewpoints. Obtaining such topographic data from one viewpoint could not completely be better than other viewpoints, due to the fact that each view has their own representations and limitations. For example, top-view surveys may hardly detect some key urban features: vertical structures (e.g. building walls, alleys, and openings) and low-level structures (e.g. kerbs, pathways, and underpasses). In contrast, it could be much easier to detect these key features from side-view surveys. Emerging of digital cameras and Structure from Motion (SfM) techniques is capable of providing new opportunities to achieve high resolutions and high accuracies of topographic data from different viewpoints. In this chapter, a new side-view SfM topographic data is discussed. Land surveys are overviewed in Section 4.1. Side-view SfM data acquisition is described in Section 4.2. Reconstructing and registration processes of raw SfM data are explained in Section 4.3. Extracting key urban features from side-view SfM data is given in Section 4.4 and issues concerning side-view SfM data are discussed in Section 4.5.

4.1 Land surveying approaches

The direct measurements, e.g. using a measurement tape, have long been used in conventional land-surveying approaches. These direct measurements could be handy performed only when target objects can be accessed. Wherever the accessibilities are limited, but can still be seen, indirect measurements could be conducted much easier. For example, a conventional levelling telescope camera (Fig. 4-1) and theodolites can be applicable to determine a distance and a relative elevation of unknown. These conventional land surveying tools can be adequately used to determine the distances of the unknown when a referencing angle and distance are measured, using triangulation surveying approach.

Fig. 4-1.   An example of conventional land survey using
a levelling telescope for elevation measurements

To determine the relative elevation, a difference elevation between the known and unknown points (locations 'A¹' and 'A²', resp. in Fig. 4-2) of the Earth's surface, aka levelling measurements, can be measured by using levelling telescope camera.

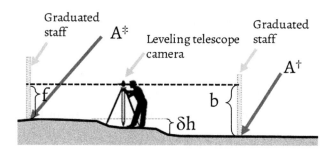

Fig. 4-2.   A conceptual graphic of levelling measurement using
a conventional levelling telescope camera

An operation of determining a difference elevation of an unknown point can be adequately used when both the unknown and known points (georeferenced point) are in line-of-sight and not too great distances from the observed location. Calculating the difference elevation can be emphasised as:

$$\delta h = b - f$$

Eq. 4-1

where: δh is the difference elevation of target point relative to the referencing point (m)

b is the measured elevation at the referencing point (m)

f is the measured elevation at the target point (m)

When conventional land surveying has shifted into a digital edge, such conventional surveying tools have turned into the more sophisticated equipment with many improvements in data quality, accuracy, and reliability. Amongst of these tools, a global positioning system (GPS; see an application of aerial surveying in Section 3.2) shows that the results are far more precise than using conventional land surveying tools.

Despite using conventional land surveying tools (e.g. levelling telescope camera) that the locations of known and unknown points need to be in line-of-sight, employing modern GPS tools can perform well without such line of sight limitations. To determine the unknown location and elevation point, a rover GPS unit and a stationary-based GPS unit (locations 'A' and 'B' in Fig. 4-3) are commonly used for synchronously recording the GPS signals from the same set of satellites. The GPS system can determine the unknown point from the signals of at least four GPS satellites. When applying the differential GPS (DGPS) approach, the recorded signals at the unknown point (location 'A†' in Fig. 4-3) can be differentiated to another known-georeferenced point (location 'B†' in Fig. 4-3), which simultaneously records the same dataset of GPS satellite signals. In post-processing approaches, such recorded GPS signals from both rover and based GPS stations are used to calculate a new absolute location and elevation of the unknown point and the unknown point can then become the known-georeferenced point as a new ground control point (GCP). The differential errors relative to the geo-referenced point can also be calculated and typically represented in the RMSE verification measure.

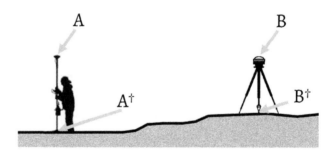

Fig. 4-3.   A conceptual graphic of ground control point measurement using a rover ('A') and a stationary ('B') based GPS units

When applying a differential GPS (DGPS) approach in land surveying, the results are far more precise (sub-metre to sub-centimetre accuracies) than such conventional surveying tools, e.g. conventional levelling telescope cameras. Even though employing stationary GPS surveys are capable of providing accurate elevation points without line-of-sight limitations or even possible to be operated at night, adopting such stationary GPS tools for measuring each GCP location require a lot of labour-intensive efforts (Aktaruzzaman & Schmidt, 2009) in the land-surveying process (measuring each GCP location using applying stationary GPS tools commonly takes 15 to 20 minutes to receive GPS signals). Reconstructing high-resolution topographic data for a small-size city, employing such DGPS approach for land surveying could still be a difficult task. Since the 1960s, emerging from laser scanning systems had substantial improvements in data quality and reliability. Not only the aerial laser scanning (ALS) system but also the terrestrial laser scanning (TLS) system has become more widely used for land surveying, aka side scanning or side-view surveying. Ever since then, the TLS systems (Fig. 4-4) have been used in the wide range applications including topography, architecture, city planning, civil engineering, mining, archaeology, etc. Despite using conventional land surveying tools, employing advanced TLS systems can acquire better topographical data in high resolution and high accuracy. Therefore, obtaining topographic data of a small-size or moderate-size city could possibly be made in a few days or weeks.

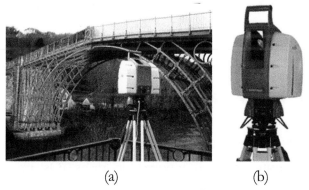

(a)   (b)

Fig. 4-4. An example of the modern terrestrial laser scanner (TLS) system:
(a) a bridge scanning application; (b) a Leica TLS ScanStation 2

Typically, side-view surveys can be used to observe target objects closer than top-view and oblique-view surveys. Therefore, side-view topographic data commonly have extreme high-resolution results with sub-metre accuracy (Fig. 4-5).

Fig. 4-5.   An example of the scanned urban features using side-view TLS tool

Even though employing side-view TLS tool can be used to substitute some missing gaps in either top-view or oblique-view topographic data, such high-resolution and high-accuracy topographic data obtained by the TLS system often come with high investment costs (Vincent, 2007). Nowadays, advances in photogrammetry and computer vision of Structure from Motion (SfM) techniques can alternatively be used to provide such high-resolution and high-accuracy topographic data by using low-cost surveying equipment. The more details of side-view SfM data acquisition are given hereafter.

4.2 Side-view SfM data acquisition

Since the mid-19th century, a stereo-photogrammetry had successfully been implemented to create analogue topographic maps (Osterman, 2007). The principles of stereo-photogrammetry follow a nature of human eyes, which can perceive target objects in three-dimensional (3D) scenes. The overlapping viewpoints seen by both eyes can be perceived as a depth perception in the brain. The furthest distance of the depth perception is limited to about 400 m because the relatively close spacing of human eyes is 6 to 7 cm apart (Drury, 1987). By adopting such principle concept, overlapping photos could also be visualised and analysed in the 3D depth perceptions by using a stereoscope equipment. Such overlapping photos could be taken from widely separated positions, but they should maintain a large number of overlapping areas between consecutive photos. However, these classical stereo-photogrammetry approaches are still time-consuming to achieve detail elevation maps.

Nowadays, Structure from Motion (SfM) techniques could be counted as a modern stereo-photogrammetry approach to create a digital topographic map, aka 3D reconstructions. Since the advances in computer vision and photogrammetry techniques have dramatically been improved, this SfM technique can provide an opportunistically accessing to the accurate topographic data in sub-metre resolutions (Fewtrell et al., 2011; Sampson et al., 2012). Despite using the sophisticated and expensive surveying tools (e.g. terrestrial laser scanner – TLS), employing normal cameras for SfM observations can adequately provide topographic data in high resolution and high accuracy (Goesele et al., 2007). Their results showed that the SfM point cloud (Fig. 4-6a) and TLS point cloud (Fig. 4-6b) are matched over 90% of both matched point clouds in pink and with 0.128 m differences of mismatched point cloud in blue (Fig. 4-6c).

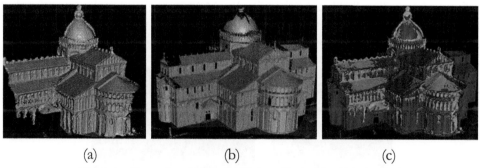

(a)   (b)   (c)

Fig. 4-6.   Examples of St. Peter, Vatican City in 3D point cloud of
(a) SfM data and (b) TLS data. (c) The comparison between
SfM and the TLS data (source by Goesele et al., 2007)

Moreover, emerging of these SfM techniques can provide the digital topographic data with the flexible and economical way (Fraser, 1997). Unlike other modern technologies which expensive and sophisticated acquisition tools, e.g. laser scanners and digital big-format cameras, low-cost acquisition tools (consumer-grade digital cameras, camcorders, or even mobile phone cameras can be adequately used for 3D reconstructions in the SfM technique. Hence, supreme resolution digital big-format cameras may not necessarily need, but consumer digital cameras should have more than 3-megapixel resolution (McCoy et al., 2014).

Many digital cameras and mobile phone cameras these days can capture photo shots and recording video scenes. Amongst of these cameras, digital single lens reflex (DSLR) cameras have more flexibilities and reliabilities than other digital camera types. Such DSLR cameras are capable of interchangeable lenses and support a vast of professional gears, e.g. flashes, tripods, and stabilisers. A larger image sensor, a better battery life, an external memory-storage support, and a more solid camera body (with or without waterproof capabilities) can be commonly found in several DSLR cameras. Furthermore, these relatively small DSLR camera bodies can be simply mounted on cars (Fig. 4-7), mopeds, or even used as handheld devices, which are easy to operate in the field. Even though the DSLR camera is often easy for handheld operation, it should be more practical to mount the camera(s) on a mobile unit, e.g. bicycle, and vehicle.

(a)   (b)

Fig. 4-7.   An example of (a) a simplified mobile unit mounted with
(b) the dual DSLR cameras of location 'A' in Fig. 4-7a

The more advances in such mobile unit can equip with other surveying tools. The more complex mobile unit could also be equipped with the more expensive and sophisticated surveying instruments. While some advanced mobile units can have dual GPS antennas, a secondary GPS antenna can be separately installed to aid in the heading determination, whereas other mobile units have orientation data obtained from IMU system. When referring all sensor antennas relative to a centroid of the IMU unit, such orientation data can be used to compensate the missing position of the mobile unit when the GPS signal are weak. Thus, the determination of these GPS and IMU units can provide the precise position of the mobile unit with sub metre accuracy. It could also be practical to apply these accurate mobile-unit positions to calculate the surface of street elevations when a mobile unit height relative to the centroid of the IMU unit is known. In this research, an advanced mobile unit was equipped with separation antennas of GPS, IMU, and dual DSLR cameras (Fig. 4-8).

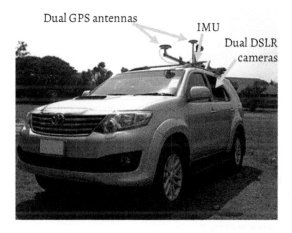

Fig. 4-8. An example of a mobile unit equipped with GPS and IMU systems to obtain ground survey points (GSPs)

In this research, a new side-view SfM data acquisition is explained in three steps: a side-view SfM data acquisition; an image pre-processing; a raw SfM data processing and registration, as given hereafter (Fig. 4-9).

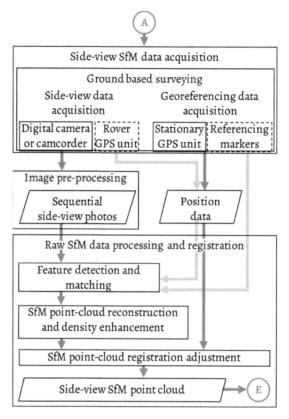

Fig. 4-9. A workflow of side-view SfM topographic data acquisition (location 'A' referred from Fig. 3-3) and further processing steps of side-view SfM point cloud are also given (description 'E' in Section 4.4)

4.3 Raw SfM data processing and registration

4.3.1 Image pre-processing

Whenever the digital camcorders are performed, the recorded videos need to be sampled into a series of sequential still photos. In-house Python scripts can be appropriately used for automated slicing a video scene into a series of overlapping photos. Some blurry scenes due to camera motions or object movements can be avoided by setting up the camera with high shutter speed, e.g. 1/800 sec, yet some moving object (Fig. 4-10) can be motley in the captured scenes. These irrelevant photos also consume a lot of disk space and computational cost. Therefore, subsequent removals of such blurred and moving-object photo scenes are needed to be undertaken manually.

Fig. 4-10. An example of a moving object is captured in the scene

Then, the selected sequential photos can be converted into the same digital file format, e.g. JPEG, TIFF, etc., with relatively the same resolution. Each file format has its unique benefit and they still have many debates about benefits of adopting these different file formats for each application. In this respect, the JPEG file is probably the most popular and widely compatible digital format supported by a number of the digital cameras and imaging software. The JPEG uses lossy compression, allowing the file can be compressed which renounce some details. A number of compressions can be varied. Due to the fact that the more compression could discard the more data, and resulting in a smaller output file. In this research, the JPEG (*.JPG) file format with no compression was used.

Masking could be used to select some uninterested areas; such masked areas of photos are not taken into account for 3D reconstruction processing. Masking can be used to reduce the result complexities by eliminating the masked areas on the photos that are not of interest. In this research, even though masking sup-steps may be useful for more precise camera positioning, such uninterested features are still needed to be selected and masked manually for each photo. These sub-steps need a lot of labour intensive efforts and time-consuming processes. Therefore, the masking was not included into these pre-processing steps.

4.3.2 Feature detection and matching

When the selected photos have been converted and stored as a JPEG file format, the key aspect of the SfM technique can then be used to determine the matching 3D positions of features in the overlapping 2D photos. Firstly, the distinct features in each photo are identified by using feature detection methods. For feature detection, two of the most popularly used methods are scale invariant feature transform (SIFT; Lowe, 1999), and the SIFT embedded with advances in multi-core graphics processing unit (SiftGPU; Wu, 2007).

In both methods, they use the two sub-steps, i.e. a feature detection and a feature description. In feature detection, the detected features (aka points of interest or key points) can be detected and identified automatically by transforming local image gradient into a representation. These features have insensitive to variation in illumination and orientation (Lowe, 2004). These key points are invariant for scaling, rotating, and partially changing in illumination of each image. The good detector needs are repeatability and reliability characteristics.

While repeatability means that the same feature can be detected in different images, reliability means that the detected point should be distinctive enough though the number of its matching candidates should be small. A number of key points are primarily dependent on the textures and resolutions for each photo. The amount of key points can be varied; the more photo resolutions, the more key points can be detected. However, some complex textures can return the better-identified key points than others (Westoby et al., 2012).

The latter is the feature description sub-step. A descriptive key point is needed in order to find corresponding pairs in the same key-point location. A descriptor is a process that extracts essential information of each identified key point, which is usually presented in the form of the point (Fig. 4-11), line, and polygon vectors. These descriptors are typically unique enough to allow these descriptive key points to be matched in large datasets.

(a)   (b)

Fig. 4-11. Decomposition using SIFT algorithm for (a) a given photos. (b) Lines represent the individual keypoint descriptor

For the matching step, when such descriptive key points are assigned, they are ready for matching with its neighbour overlapping photos. They should also have invariant to rotation, scaling, and affine transformation. To reduce the number of possible matches on the different photos, the same descriptive key point should be characterised by almost the same distinctive value.

These key points in consecutive photos can be matched using approximate nearest neighbour (Edelman et al., 1998) and Random Sample Consensus (RANSAC) algorithms (Fischler & Bolles, 1981). A track linking specific key points in a set of overlapping photos can then be established. These tracks comprising a minimum of two key points and three images were maintained for the next steps (Fig. 4-12), with those which fail to meet these criteria being automatically removed (Snavely et al., 2006) by using visibility and regularisation constraints (Furukawa & Ponce,

2010). By using these methods, transient features such as moving objects are discarded from the dataset, due to they are not suitable for employing in SfM reconstructions and their relative positions to other key points are constantly changing.

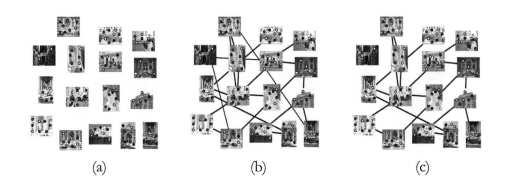

(a)   (b)   (c)

Fig. 4-12. Processes of (a) feature detection, (b) feature matching, and (c) feature matching adjustment (source by Snavely et al., 2006)

The complex variations, e.g. lighting, material textures, etc., may also influence quality of taken photos. It seems to be not possible to offer explicit guidance on a minimum number of photo shots, which are necessary for a successful SfM reconstruction. Theoretically, three overlapping photos taken from different viewpoints are required as a minimum. It is not as always to capture a huge number of photo scenes, as well. However, obtaining five to ten overlapping photos is feasible recommended to optimise the ultimate number of keypoint matches and system redundancy, with a minimal logistical constraint.

4.3.3 SfM point cloud reconstruction and point cloud density enhancement

To reconstruct the 3D data from overlapping 2D photos, the implicit parameters and explicit camera orientations are valuable information. These parameters can be next used to attempt to recover camera positions from correlative rotations, projections, and transformations of corresponding photo scenes by using 3D geometrical calculations. Following the matching step, the sparse bundle adjustment system (Snavely et al., 2007; Wu et al., 2011) is used to estimate camera pose. It also used to improve the accuracy, minimise projection errors from camera tracking, and refined the camera positions in order to extract a low-density or 'sparse' SfM point cloud data. In this stage, keypoint correspondences place constraints on camera pose orientation, which are reconstructed using a similarity transformation, whereas minimising errors is achieved by using a nonlinear least squares solution (Szeliski & Kang, 1994; Nocedal & Wright, 1999). Finally, triangulation is used to estimate the 3D point positions and incrementally reconstruct scene geometry, fixed into a relative coordinate system.

From the previous step, sparse SfM point cloud data are very discrete and neither enough for analysing nor visualisation. The enhancements on these sparse point clouds (Fig. 4-13b) are needed to make a denser SfM point cloud. An enhanced density point cloud can be derived by implementing the Clustering View for Multi-View Stereo (CMVS) and Patch-based Multi-View Stereo (PMVS) algorithms (Furukawa & Ponce, 2010).

The camera positions derived from Bundler are used as input. CMVS then decomposes overlapping input images into subsets or clusters of manageable size, while PMVS2 is independently used to reconstruct 3D data from these individual clusters. The result of this additional processing is a significant increase in point cloud density (Fig. 4-13c). Some software package, e.g. PhotoScan™, can generate the point cloud in different ranges of density. Obviously, the more point cloud density, the more computational time will be used.

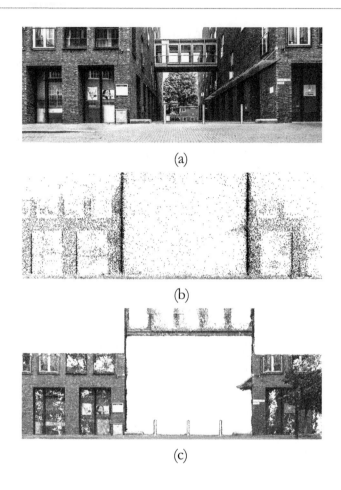

(a)

(b)

(c)

Fig. 4-13. As example of a high-density SfM point cloud represents
the connecting structure between two buildings

4.3.4 SfM point cloud registration adjustment

Between the consecutive SfM point cloud data sets, when the correspondences of
two or more strips of the dataset were required, they need to be carefully defined
in the same georeferencing system, e.g. world geodetic system 1984 (WGS 1984 or
EPSG: 4326). Due to a lot of SfM point cloud strips were independently generated
into several different locations, obtaining the GCPs for each strip were labour
intensive and time-consuming approaches. Therefore, some GCPs can be used for

some SfM strips. Therefore, applying some identical urban features or referencing markers, which appears in the overlapping area for the rest of SfM strip, should be appropriately used as referencing GCPs. These referencing markers should have sharp contrasts with the surroundings, and they should be placed to provide maximum visibility (Fig. 4-14). However, such integrations are only possible when the reference objects can be visible in both the consecutive point cloud data sets. Another approach is to employ the registered top-view LiDAR data as geo-referencing points. Therefore, the systematic errors can be minimised compared to the previous approach.

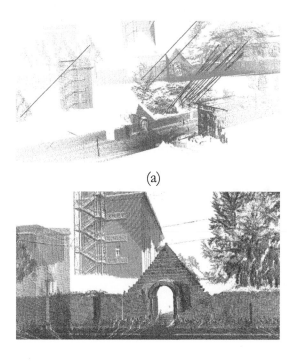

(a)

(b)

Fig. 4-14. (a) Referencing markers correlated in between two consecutive side-view point cloud; (b) final result of the point cloud adjustment (source by PCL)

To check the effectiveness of the registered LiDAR data, another set of GCPs obtained from the GPS system can be performed by using three cross-verification measures (see examples in Section 3.3; Knotters & Bierkens, 2001): (i) the systematic error or mean (ME), (ii) coefficient of determination (R^2), and (iii) the root mean squared error (RMSE). Residual errors from georeferencing of the GCP points can consider being accurately measured with sub-metre errors.

In this research, the residual errors from georeferencing of the GCP points found to be accurate measurements with sub-metre scale errors for all case studies. Typically, the final result of side-view SfM data can have a horizontal and vertical RMSE accuracy of $\pm$ 0.10 m. Due to side-view SfM surveys observes closer to target objects, its side-view topographic data can have extreme high-resolution of 0.20 m or less.

Therefore, it may contain an extremely large number of elevation point cloud data. However, high-resolution side-view SfM data may include an extremely large number of elevation points, aka the point cloud data. This point cloud can be ranging from tens of millions to billions of elevation points. This massive point cloud can cause a data storage issues. Only a single large file may be several hundreds of megabytes to several gigabytes.

4.4 Side-view SfM data simplification

When side-view SfM topographic data are obtained, essential information can be extracted from the vertical urban features. In this respect, the three distinct key components (i.e. low-level structure, façade, and opening point cloud) were classified (Fig. 4-15), as described hereafter.

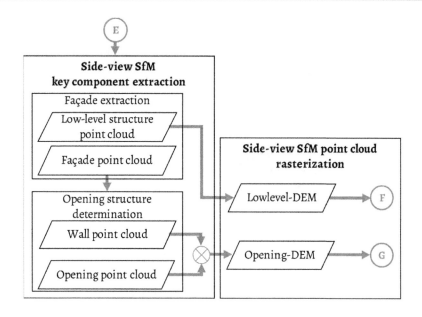

Fig. 4-15. A workflow of the key component extraction using side-view SfM point cloud (location 'E' referred from Fig. 4-9)

The further processing steps of low-level structure and opening point cloud are described hereafter. All these extracted low-level structures, wall, and opening point cloud are rasterised as Lowlevel-DEM and Opening-DEM (locations 'F' and 'G', resp. in Fig. 4-15). Also, such façade point cloud data are used for flood watermark extraction in Section 7.4.

4.4.1 Façade and low-level structure point cloud extractions

A side-view SfM point cloud may contain many elevation points representing the different urban features. Although main benefit of adopting side-view SfM data is for substitute some missing gap from conventional top-view data, some key components of small urban features can also be gained by product. In this research, low-level structures, façades, and opening walls were three main interesting urban features. Firstly, a buffer boundary of a building footprint (location 'A' in Fig. 4-16) was enlarged with buffering area of ±0.5 m (location 'B' in Fig. 4-16). This buffer area should cover side-view SfM point cloud. When such SfM points appeared into this buffer area, they were then classified as a façade point cloud. This façade point cloud contains walls and building structures.

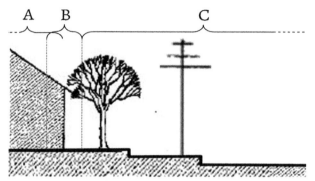

Fig. 4-16. A conceptual map schematic profile view of building different boundaries relative to the building ground footprint

This simple algorithm can be chosen when the building footprint boundaries are given or even created by analysing extracted building point cloud from conventional top-view LiDAR data (Section 3.4). When using given boundaries, the building structures (location 'A' in Fig. 4-17) and façade structures (location 'B' in Fig. 4-17) can be extracted from surrounding urban features. Whereas the rest areas out of building and buffered boundaries were classed as mixed surrounding urban features, e.g. kerbs, trees, poles, and roads (location 'C' in Fig. 4-17).

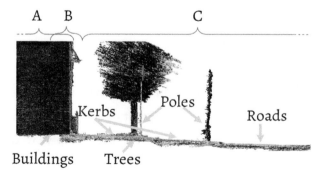

Fig. 4-17. An example of side-view SfM point cloud represented in a profile-view projection

Such mixed feature point cloud can then be re-segmented once again. The point cloud lower than a threshold relative to the given ground elevation were selected and kept as low-level structure point cloud.

In particular, these extracted low-level structure point cloud (location 'C1' in Fig. 4-18) maintains kerbs and roads with tree stems and based poles, but any top of sidewalk trees and electric poles point clouds (location 'C2' in Fig. 4-18) are all removed.

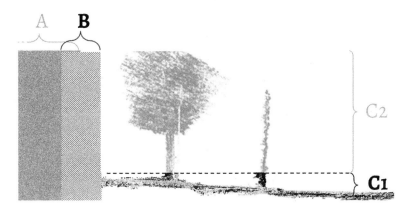

Fig. 4-18. An example of low-level structures point cloud represented in a plan view projection

Despite using the sophisticated algorithms, extracting essences of these key features by applying given boundaries of building footprints could be adequately used for extract surrounding urban features from building structures.

In more details, some small urban features (e.g. kerbs, parking lanes, and roads; Fig. 4-19) could have a considerable effect on floodwater dynamics and predictions (Remondino & El-Hakim, 2006; Rychkov et al., 2012; Sampson et al., 2012). Even though filtering algorithms could help to remove most obstacles of urban features, they often remove all small urban features.

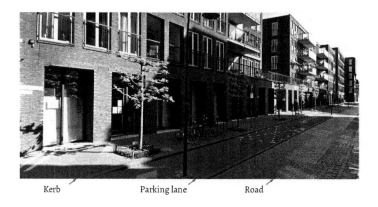

Fig. 4-19. A photo scene example represents low-level structures

Due to the fact that the filtered elevations (Fig. 4-20b) were not considered with urban features (i.e. kerbs, parking lanes, and roads; location 'Xa' in Fig. 4-20a).

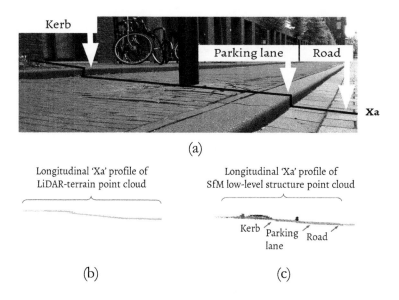

Fig. 4-20. A zoomed-in detail of (a) a photo scene of low-level structures with (b) a top-view LiDAR flat terrains and (c) a side-view SfM low-level structures along the 'Xa' cross section

While such high urban features and small urban features (sometimes hidden underneath high features) were filtered out and not remained in the LiDAR-DTM, in contrast, such small urban features were still maintained in new MSV-DEM. Owing to this, a side-view SfM topographic data can better represent such small urban features from extracted low-level structures (Fig. 4-20c). Some essentials of these small urban features could be translated such drainage capabilities from roads and pathways to enhance 2D model schematics (Sampson et al., 2012).

A new MSV topographic data (in combination of side-view SfM data with top-view LiDAR data) contains more key components of complex urban features than conventional single-source view data. Many researchers (Haile & Rientjes, 2005b; Hunter et al., 2008a; Boonya-aroonnet, 2010; Razafison et al., 2012) noted that details of such key components (e.g. alleyways, passages, and kerbs) could have a major impact on flood propagations and flow dynamics in 2D flood simulation results.

In this respect, MSV-DEM could be adequately used as topographic input for enhancing 2D flood schematics. The key components in this new DEM could result in a more realistic representation of inundation depths and floodwater dynamics for those complex flood situations.

Moreover, when the study area is relatively flat, a given referencing elevation and maximum-minimum threshold elevations should be adequately used to classify low-level structures. Such elevation points matched within this threshold can then be extracted and classed as low-level structure point cloud. A conceptual algorithm of low-level-structure point cloud extraction is presented, as follow (ALG. 4-1).

ALG. 4-1. A conceptual algorithm for side-view SfM low-level-structure extraction

DATA Registered side-view SfM points, rawSfM
BEGIN
    SET Initial terrain elevation, iniTerrain //e.g. 5 m msl
    SET Terrain threshold maximum, thresholdMax //e.g. +0.5 m referred to
iniTerrain
    SET Terrain threshold minimum, thresholdMin //e.g. -0.5 m referred to iniTerrain
    SET Buffer area of building footprint, buildingBuff //e.g. ±0.5 m
    SET Selected SfM-facade points, façadeSfM = ""
    SET Selected SfM non-facade points, non-façadeSfM = ""
    SET Selected SfM low-level points, lowlevelSfM = ""
    SET Key feature field to all points, keyFeature = ""
    SET Projection plain, projPlain = z
    PROJECT rawSfM-points along projPlain
    WHILE projPlain = z;
        IF rawSfM -points in buildingBuff
            KEEP rawSfM-points, façadeSfM
            ASSIGN keyFeature = "façade", façadeSfM
        ELSE
            KEEP rawSfM-points, non-façadeSfM
            ASSIGN keyFeature = "non-façade", non-façadeSfM
        END IF
    END WHILE
    SET Projection plain, projPlain = z
    PROJECT non-façadeSfM-points along projPlain
    WHILE projPlain = z;
        IF non-façadeSfM-points in (iniTerrain AND
        thresholdMax AND thresholdMin)
            KEEP non-façadeSfM-points, lowlevelSfM
            ASSIGN keyFeature = "lowlevel", lowlevelSfM
        END IF
    END WHILE
END

For further terrain analyses (Section 5.3), the extracted lowlevelSfM point cloud data (Fig. 4-21a) were then used to create a Lowlevel-DEM (Fig. 4-21b).

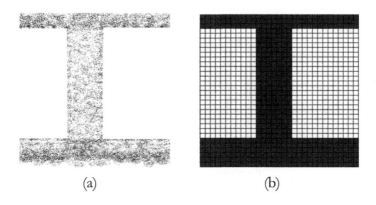

(a)   (b)

Fig. 4-21. Two representation examples of extracted low-level features: (a) a lowlevelSfM point cloud; (b) a raster-based Lowlevel-DEM at 1 m grid resolution

However, some manual selection still needed to be performed for maintaining some connecting structures and walls while removing standing peoples and unrelated features. This simple and straightforward approach was chosen due to, all case studies are relatively small.

4.4.2 Determination of openings around structures

The selected façade point cloud still contains with closed walls and hidden openings. Such hidden openings may hardly be detected from the sky. When using side-view SfM point cloud, hidden openings, e.g. pathways or underpasses under overarching structures (location 'A' in Fig. 4-22), openings under trees (location 'B' in Fig. 4-22), and roads under sky train tracks (location 'C' in Fig. 4-22) can be easier for identifying from side viewpoints. Such hidden openings commonly have a pathway underneath them.

In side-view SfM point cloud, above this pathway may contain a sparse point cloud and generally represent an empty open space (location 'A' in Fig. 4-22).

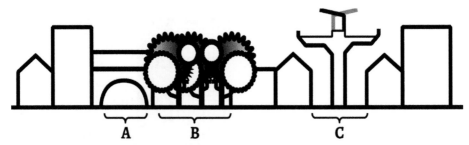

Fig. 4-22. A schematic of side-view surveying that can be easy to identify the major opening structures

In this research, a straightforward approach can be adequate to identifying such empty open space (hidden openings) from surrounding façades. A simple moving window algorithm was used for opening detections, whenever such façades are relatively flat and small. For example, a moving window of 0.25 m² was applied which would be appropriate to identify such open space and therefore available for translating them into a high-resolution model schematics. This moving window will be moving along a façade planar.

In this example, the threshold of point cloud data was set at the minimum of 2,000 points for classifying close walls and opening spaces. When examining the number of points in the moving windows was lower than the threshold, that window block will be defined as an opening-space block (mark 'O' in Fig. 4-23b). Whereas a closed-wall block (mark 'W' in Fig. 4-23b) will be defined when the number of points was higher than the threshold. Examining window will continuously move according to all rows and columns until they are no more points for each observing façade planar.

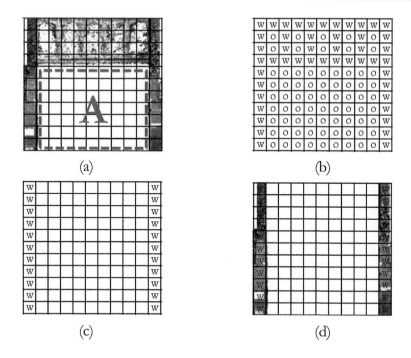

Fig. 4-23. Conceptual diagrams of applying moving windows for
determining hidden openings

Along these planar, whenever opening blocks were found in between wall blocks along the same column, the lowest wall block of such column will be maintained and declared as a confirmed opening block (mark 'O' in Fig. 4-23c). All continuity wall blocks remained the same but all opening blocks will be marked as empty blocks in Fig. 4-23c). Wall and opening blocks were thereafter used for classifying side-view SfM point cloud (Fig. 4-23c).

When crest heights of openings occurred, they could also be maintained for representing the actual elevation heights of such opening crests, which may not always have the same elevation of their surrounding terrains. The mixed closed walls and hidden openings can be extracted by using a moving window algorithm (ALG. 4-2) applied as a simple opening extraction for side-view SfM data.

ALG. 4-2. A conceptual algorithm for side-view SfM opening extraction

DATA Selected SfM facade points, façadeSfM
BEGIN
    SET Counted points in each moving window, countedPoint = ''
    SET Moving window size, windowBlock //e.g. 0.25 m^2
    SET Marked value to each moved window, markedBlock = ""
    SET Selected SfM-wall points, wallSfM = ""
    SET Selected SfM-opening points, openingSfM = ""
    SET Key feature field to all points, keyFeature = ""
    SET Projection plain, projPlain = x OR y
    PROJECT façadeSfM-points along projPlain
    WHILE projPlain = x OR y;
        COUNT (rows, columns) along projPlain
        FOR each (row, column) do
            COUNT façadeSfM-points in
            windowBlock-criteria, countedPoint
            IF countedPoint > 2,000
                ASSIGN "W", markedBlock
            ELSE
                ASSIGN "O", markedBlock
            END IF
        END FOR
        ...
        FOR each (row, column) do
            IF markedBlock = "W"
                KEEP façadeSfM-points in markedBlock
                ASSIGN keyFeature = "wall", wallSfM
            ELSE
                REMOVE the remaining façadeSfM-points
            END IF
        END FOR
    END WHILE
END

From given example, a moving window of 0.25 m² should be applied for a high-resolution model schematics (i.e. 1 m grid resolution). It could be adjustable for a coarse-resolution schematic with four to one ratio, when a (four-time) higher resolution of extracted point cloud should be adequately represented and translated essences of key features in a (one-time) coarser resolution of simplified 2D model schematics. Again, the density threshold of point cloud could also be adjustable according to the moving window size and the density of original point cloud.

Furthermore, some key components of small structures (e.g. doors, windows, and walls made from glass) could be further analysed and extracted. During the flood event, these closed structures may be potentially easy to break (breakable structures). Such susceptible opening cells (Fig. 4-24) are prone to flooding.

Fig. 4-24. An example of floodwater through the susceptible doors
(source by Virginia Living Museum)

These breakable structures are often neglected in conventional land surveying. In SfM 3D reconstructions, urban structures with glassy, shiny, and reflective surfaces could have a low point cloud densities (Fig. 4-25b). These sparse point cloud data have to be avoided for most SfM 3D reconstructions, but such limitation was not always the case for this research. In contrast, that sparse point cloud with discontinuity surfaces can give an opportunity for identifying such glass structures as potential openings, which concealed into the façades.

(a)   (b)

Fig. 4-25. A side view of a mixed glass and brick façade: (a) a photo scene; (b) a high-density SfM point cloud reconstructed by side-view SfM technique

Again, all mixed closed walls and susceptible openings in façade point cloud can be extracted by using the simple opening extraction algorithms (ALG. 4-2). Such simple algorithm of a moving window was used to determine such susceptible opening structures (e.g. glass wall, glass windows, and glass doors).

When a moving window of 0.25 m² was defined and it will then be moving along a façade planar. The threshold of the point cloud was defined. Thereafter examining the number of points in the moving windows was lower than the threshold that window block will be indicated as a susceptible opening block (mark 'O' in Fig. 4-26b). Whereas a wall block (mark 'W' in Fig. 4-26b) will be defined when the number of points was higher than the threshold. Examining window will continuously move according to all rows and columns until they were no more points for each observing façade plain.

However, when the façadeSfM point clouds are too complex, human interpretations are recommended for determining and extracting such complex openings.

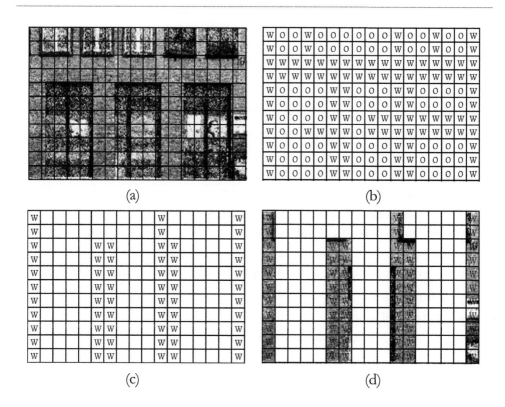

Fig. 4-26. Conceptual diagrams of applying moving windows for
determining susceptible openings

4.4.3 Side-view SfM mapping and rasterization

In this section, the extracted side-view SfM point cloud was used to create a digital
elevation model (DEM), employing the IDW algorithms in the rasterization
process. Concerning complexities of (hidden) urban features for a city is crucial.

Therefore, the extracted low-level structures and vertical structures (i.e. walls and
opening pathways) point clouds of side-view SfM data were used to create the
raster-based SfM-DEM at different resolutions. A conceptual algorithm for side-
view SfM-DEM rasterization is presented (ALG. 4-3).

ALG. 4-3. A conceptual algorithm for side-view SfM-DEM rasterization

DATA Extracted side-view SfM points, lowlevelSfM, wallSfM, openingSfM
BEGIN

    SET Raster grid size, sizeGrid //e.g. 1 m, 5 m, 10 m, or 20 m grid resolutions

    SET Sampling criteria, samplingPoint = maximum

    SET Sampled point value, sampledValue = ""

    SET Assigned DEM-lowlevel grid, lowlevelGrid = ""

    SET Assigned DEM-wall grid, wallGrid = ""

    SET Assigned DEM-opening grid, openingGrid = ""

    SET Assigned DEM-null grid, nullGrid = ""

    SET Projection plain, projPlain = z

    PROJECT (lowlevelSfM-points AND wallSfM-points AND openingSfM-points) along projPlain

    WHILE projPlain = z;

        COUNT (rows, columns) along projPlain

        FOR each (row, column) do

            IF active cell at (row, column) contain

            keyFeature = "lowlevel" as majority

                SAMPLING (lowlevelSfM-points) with samplingPoint-criteria, sampledValue

                ASSIGN-GRID by interpolating sampledValue with sizeGrid-criteria, lowlevelGrid

                ...

            END IF

        END FOR

        FOR each (row, column) do

            CREATE-DEM use (lowlevelGrid, nullGrid) with sizeGrid-criteria, Lowlevel-DEM

            CREATE-DEM use (wallGrid, openingGrid, nullGrid) with sizeGrid-criteria, Opening-DEM

        END FOR

    END WHILE

END

For further terrain analyses (Section 5.3), the extracted wall and opening point clouds (Fig. 4-27a) are later used to create the Opening-DEMs (Fig. 4-27b). In this respect, all essences of key features (i.e. low-level structures, walls, and openings) should be adequately represented and translated to the final resolutions of raster-based SfM DEMs, i.e. Lowlevel-DEM (Fig. 4-21b).

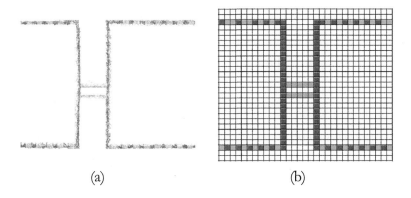

(a)   (b)

Fig. 4-27. Two representation examples of extracted wall and opening features: (a) a wall and opening point cloud; (b) a raster-based Opening-DEM at 1 m grid resolution

4.5 Issue concerning the side-view SfM data

Overall errors of different-source views topographic data can obviously be accumulated from both the data acquisition and data processing stages. For side-view SfM topographic data, errors in data acquisition stage can be caused by camera tools and their setups. Low-quality photos (e.g. low resolution, low details, burly photos, and noisy photos) can degrade an output quality of SfM reconstruction results. Insufficient overlapping between photos can cause an unsuccessful process in SfM reconstruction. A large overlapping area in consecutive photo scenes is more important than a large number of photo shots.

Even though capturing photos from different views as many images as possible may guarantee a success of SfM reconstruction results, adopting such approach for a large study area could also come up with a large number of photo collections. Whenever the number of photo shots is increased, computational times in SfM reconstructions are exponentially growing (Golparvar-Fard et al., 2011). Despite taking many photo scenes from the same view angles with the same zoom settings, shooting photos from different perspectives, different viewpoints, and different zoom settings are highly recommended. Each photo shot should capture the same object(s) at least from three different viewpoints (McCoy et al., 2014). Overlapping areas of these consecutive photos should have an angular difference between 25 to 30 degrees. (Ducke et al., 2011). However, capturing overlapped areas for such consecutive photos can become more difficult, especially for complex urban areas.

In data processing stage, errors can be caused by geo-positioning, orientation alignment, and data integration processes. Whenever topographic data (i.e. top-view LiDAR data or side-view SfM data) are integrated, errors can also be accumulated from (sub) processes for each different-source view data. However, for side-view SfM topographic data, such errors can be minimised when data (pre) processing stage is applied for all taken photo shots. In this respect, objects in motions, glare and glow effects (Jadidi et al., 2015) including reflective and shiny objects should be avoided. Also, cropped photos should not be used due to they can lead to incorrect location estimations in SfM reconstruction results. Avoiding and excluding such low-quality photos in such data pre-processing stage could improve quality of side-view SfM data. Furthermore, minimising distances between the camera and the object(s) should also provide more quality reconstruction results (Xiao et al., 2008).

Some researchers (Baltsavias, 1999; Dai & Lu, 2010; Dai et al., 2013; Dai et al., 2014) reviewed and evaluated benefits of employing SfM techniques for construction surveying and their related applications. However, adopting SfM technique is not ubiquitous for some 3D reconstruction scenarios due to some systematic errors. For implicit camera factors, such systematic errors could be minimised by applying the following recommendations. For example, a radial distortion may lead to deviations from the central perspective model; such symmetric radial distortion can be defined as the radial displacement of an off-axis target either closer to or further from principle point. Such implicit camera factors could be appropriately adjusted by using professional calibration tools before using the camera in the field. Another recommendation is that employing a large digital camera sensor whenever possible. The larger sensor could provide a better photo quality with low noise for each captured photo, especially for low-light situations. Moreover, using a digital single-lens reflex (DSLR) could have more flexibilities from interchangeable lenses. In this respect, using prime lenses generally provide better photo quality than zoom lenses.

For the systematic error of explicit camera geometry, there also have some recommendations that should follow. For example, increasing overlapping areas of each consecutive photo between 60% and 90% should have a more satisfactory rate for 3D reconstruction results. An external orientation angle of the cameras should be set perpendicular to the mobile unit direction. Apart from both systematic error sources from implicit camera factors and explicit camera geometries, these SfM 3D reconstructions also suffer from difficulties to reconstruct textureless surfaces of glass walls or windows.

Even though the SfM 3D reconstruction is more portable and low cost (in surveying equipment, and low-cost or freely available software products, e.g. VisualSFM), the data-ready approach such as TLS system needs less time in data processing. In SfM reconstruction, feature detection and matching; sparse and dense reconstruction algorithms; are computationally demanding. However, the computational power is nowadays reducing run times for SfM processing. Moreover, side-view SfM point cloud can be excess of $\sim10^6$ of points in 10 m². It needs more data storage and more data handling. Almost all surveying technologies are limited by line-of-sight of the sensors to the targets. Certainly, side-view SfM data acquisition has also no exception. On the ground, some missing data of the enclosed terrains placed behind walls or connecting buildings can be easily substituted by top-view data. On the contrary, some missing data of the vertical objects placed behind other structures can be hardly detected either top-view or side-view surveys. In fact, the SfM technique can also be applied to reconstruct the 3D point cloud for both the façade surfaces and interior spaces, as long as the target objects are in the line of sight. In other words, the accessibility to target objects certainly is the main limitation of hidden object, neither not acquisition techniques nor surveying viewpoints. Therefore, significant processing loads and choices of survey method will eventually be weighed against a number of factors, including cost, accessibility, experience, and fitness-for-purpose regarding data resolution and coverage (Westoby et al., 2012). In this research, side-view SfM data acquisition mainly concerns to the exterior façade, which can be detected from streets.

CHAPTER 5
A novel approach for merging multi-views topographic data

Topographic data are commonly obtained from a single viewpoint (e.g. top-view LiDAR data). Because of this some key urban features may not be captured. When topographic data are merged with side view data, a novel approach of enhancement by multi views is introduced in Section 5.1. Effects of grid size and considerations for raster-based topographic data are further explored and discussed in Section 5.2 and 5.3, resp. Selection of case study areas is given in Section 5.4.

5.1 Multi-view enhancements

5.1.1 Top-view LiDAR data

Even though aerial LiDAR surveys can be used to acquire top-view LiDAR topographic data, which have long been used in many applications, in complex cities (Fig. 5-1) these conventional top-view LiDAR data may have difficulties to represent some key components of urban features, e.g. building façades, walls, alleys, opening pathways, and other urban features.

Fig. 5-1. A 3D view example of complex urban features
in Delft (Google Earth™ 7.1.5.1557, 2017)

The top-view LiDAR data can only represent a rooftop of a walkway as an overarching structure (location 'A' in Fig. 5-1; Fig. 5-2).

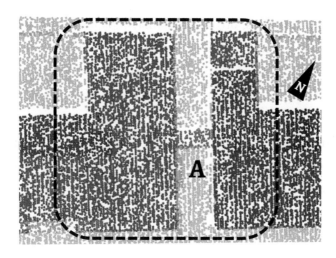

Fig. 5-2. An example of an overarching structure represented in top-view LiDAR
point cloud data (area 'A' in Fig. 5-1; AHN 2 (2014))

In LiDAR-DSMs, buildings and overarching structures (e.g. the connected
walkway between two buildings, location 'A' in Fig. 5-3b) could be seen as obstacles
in 2D models.

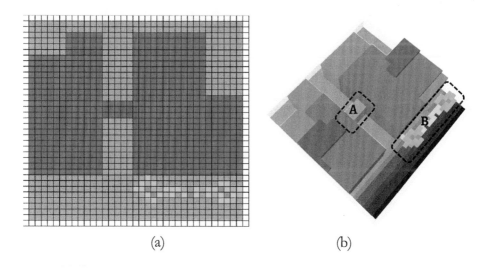

(a)   (b)

Fig. 5-3.   Two examples of 1 m grid resolution LiDAR-DSM:
(a) a plan view and (b) a perspective view

In dynamic flood situations, urban floods may rapidly develop during the initial period of the flood event. Key components of extracted urban features in raster-based topographic data should be carefully prepared and properly translated into flood model schematisations. Spatial heterogeneities of such key components in model schematisation (i.e. walls and opening pathways) can play a significant role in flood dynamics. Such key components of either natural or manmade features could behave as pathways or obstacles in flood simulation. Leandro et al. (2016) showed that complex key features can have a medium to high impact on flood depths.

In the Delft case, simulated flood evolutions using LiDAR-DSM showed that floodwaters cannot propagate further to the North (Fig. 5-4b), since the overarching sky bridge was perceived as an obstruction (location 'A' in Fig. 5-4a).

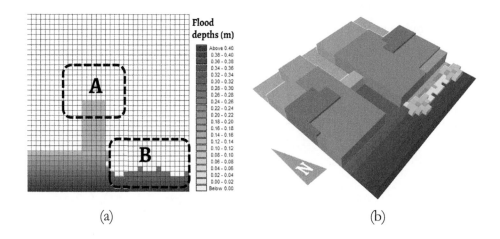

(a)   (b)

Fig. 5-4.   Examples of simulated flood using LiDAR-DSM as input visualised at the final simulation time for two different views: (a) a plan view flood-depth map with scales on the right and (b) a perspective view flood-depth map

Even though top-view LiDAR data were reconstructed at very-high resolution, applying such high-resolution data may not provide an accurate result. Although higher temporal and spatial resolutions commonly increase computational costs for simulations, they may not always provide a better flood simulation result. A way to get better representation of complex urban features should rely on adequate qualitative analysis of topographic data obtained from other sources and different viewpoints.

5.1.2 Side-view SfM data

Unlike top-view surveys, hidden urban features can be observed much easier from different view surveys. Observing urban scenes from side-view surveys allows representing openings and pathways (Fig. 5-5), which are commonly hidden underneath overarching structures.

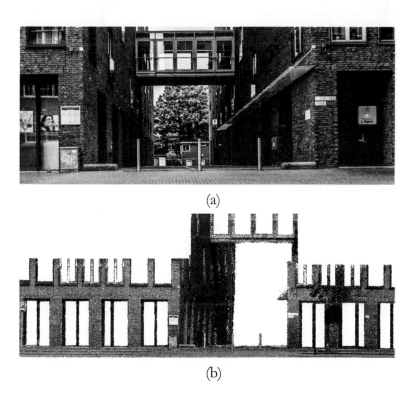

(a)

(b)

Fig. 5-5. Two examples of an overarching structure represented in
(a) a side view photo and (b) a side-view SfM point cloud data

When applying LiDAR-DSM as input, such overarching structures are commonly represented as building rooftops and seemingly behave as an obstacle in 2D model simulations, whereas high trees can be perceived as high buildings (location 'B' in Fig. 5-3b; Fig. 5-6).

Fig. 5-6. A photo scene example represents high trees (location 'B' in Fig. 5-3a)

5.1.3 Multi-views data

A new concept of multi-source views (MSV) topographic data was applied for fusing top-view LiDAR data with side-view SfM data as input for 2D model schematics. A simple approach is to preserve the merged point cloud data and eliminate irrelevant point cloud data above a defined elevation height. Such defined elevation should be low enough to expose hidden openings and pathways, but it should be high enough to alleviate floodwater to reach that level. By merging the side-view SfM data (Fig. 5-7a), such missing pathways hidden underneath overarching structures can be substituted at opening locations, in this way revealing such missing underpasses (Fig. 5-7b).

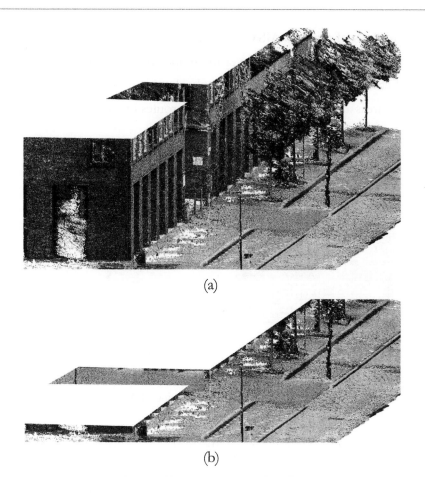

(a)

(b)

Fig. 5-7. An example of a hidden pathways represented in (a) a new MSV
point cloud and (b) an adjusted MSV point cloud

Unlike the top-view LiDAR topographic data (Section 5.1.1), the new concept of multi-source views (MSV) topographic data maintains all information of top-view and side-view point cloud data, which now can be translated for creating the novel raster-based MSV topographic data.

Since the side-view SfM surveys could give new opportunity to indicate and determine missing openings, the obstacle cells of conventional top-view LiDAR data could be substituted by terrain elevations on such opening locations. The essential features of this new MSV data can thereafter be used for enhancing 2D model schematics (Fig. 5-8). Whenever these openings are embedded into the 2D model schematics, the flood simulation results should highlight some missing flood inundation areas and reveal more sensible flow routes in complex cities.

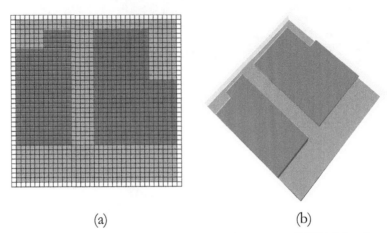

(a)   (b)

Fig. 5-8. Two representation examples of (a) a new MSV-DEM and
(b) a perspective view of 2D schematic using MSV-DEM as input

Whenever such pathways are exposed, the flood simulation results should reveal proper flood inundation areas and flood flow routes. In particular, an actual pathway hidden underneath (underpass) could play an important role in dynamics urban-flood analyses.

When applying the new MSV-DEM as input for 2D flood models, overarching structures were replaced by an opening pathway and high trees were substituted by low-level stumps (locations 'A' and 'B' in Fig. 5-3b). In this way, the simulated flood evolutions using MSV-DEM show that floodwater can now freely propagate (Fig. 5-9) through a pathway. High trees can also be filtered out (location 'B' in Fig. 5-4a compared to Fig. 5-9).

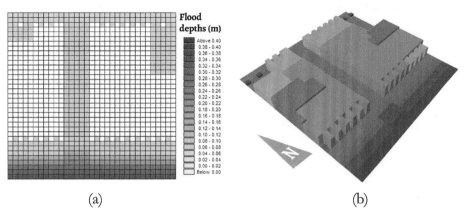

(a)   (b)

Fig. 5-9.   An example of simulated flood simulation using MSV-DEM as input visualised at
the final simulation time represented in (a) a plan view flood-depth map
with scales on the right and (b) a perspective view flood-depth map

In this research, merging multi-source view (MSV) topographic data from top-view
LiDAR data with side-view SfM data is used to enhance 2D model schematics for
complex urban-flood analyses.

Moreover, scale factors in case of grid-based resolutions for 2D schematisation
should adequately be taken into account. These scale factors crucially depend on
the required resolution of the particular flood simulation application.

5.2   Effect of grid size

Typically, a 2D schematic resolution is often coarser than the scale of urban
features. The spatial resolution of a 2D model could vary from 0.5 m to 20 m for
urban flood-model applications. While top-view LiDAR data may vary from 0.25
m to 2 m resolutions whereas side-view SfM data may vary from 0.15 m to 1 m
resolutions, such different resolutions of multi-view topographic data are
commonly translated to coarser 2D model schematics.

For example, high-resolution top-view LiDAR data are typically simplified (degraded) and translated to a 20 m grid resolution 2D model schematic. Owing to this, finer-resolution side-view SfM data (Fig. 5-10a1 to a3) could be locally identified for matching the exact location of top-view LiDAR data to create new MSV-DEMs (Fig. 5-10b1 to b3). The elevation in coarse grid cells can be obtained by averaging elevations of the existing surrounding terrain. A number of small openings could be scattered over a façade.

In this case, they could be accumulated as one big opening (Fig. 5-10b2). However, when a few small openings occurs, the modeller should consider defining the small opening (Fig. 5-10a3) as one big opening (20 m width; Fig. 5-10b3) taking into account increased roughness resistance (see Section 5.3.2.)

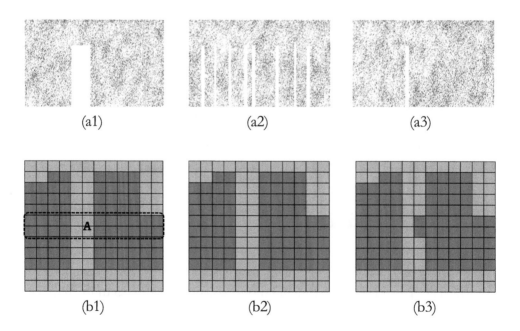

(a1) (a2) (a3)

(b1) (b2) (b3)

Fig. 5-10. Grid modification examples of using (Fig. 5-10a1 to a3) side-view SfM data for enhancing grids of (Fig. 5-10b1 to b3) MSV-DEMs

Typically, raster based data using square-matrix structured grids can these days be commonly found and are used in several topographic data products. These square structured grids are straightforward and they usually consume less time in data preparation and data processing (Fairfield & Leymarie, 1991). Whenever the extracted urban features are properly processed, they will be ready for constructing (sampling, interpolating, and creating) different types of raster-based computational grids. In this way, all essential features can be can be contained in the final raster-based topographic results.

In this section, merging multi-source views (MSV) topographic data mainly focuses on top-view LiDAR and side-view SfM data sources. Such MSV approach can substitute missing data in one viewpoint by other data obtained from another viewpoint. An overview of merging MSV processing steps is described hereafter (Fig. 5-11).

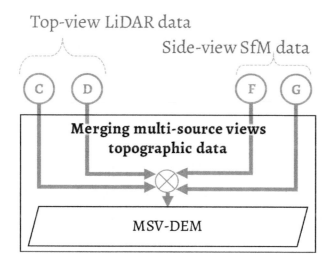

Fig. 5-11. A workflow of merging MSV topographic data using top-view LiDAR DEMs ('C' and 'D' details referred from Fig. 3-11) and side-view SfM DEMs ('F' and 'G' details referred from Fig. 4-15)

In general, all topographic data have the same georeferencing system (EPSG 4326: WGS 84). When buildings, walls, and low-level structures are located in the same cell, the elevation of the lowest component will be chosen (examples in Fig. 5-12).

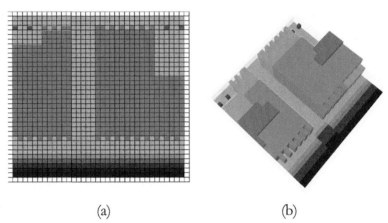

(a)   (b)

Fig. 5-12. Two representation examples of 1 m grid resolution MSV-DEM:
(a) a plan view and (b) a perspective view

Terrain and building cells will be maintained under the same conditions except when multiple small openings have been detected in the same building cells.

Such a building block (connecting building cells) can be counted as one opening pathway using terrains or low-level structures as ground level elevations of underneath overarching structures. The conceptual algorithm of MSV-DEM rasterization is presented below in ALG. 5-1. The pseudo-code was implemented in various software packages, such as Cloud Compare™ and ArcGIS®. Other choices of key features for multi-source views (MSV) grid assignments are given in Table 5-1.

ALG. 5-1. A conceptual algorithm for MSV-DEM rasterization

DATA Raster-based top-view and side-view topographic data, Building-DEM, Terrain-DEM, Lowlevel-DEM, Opening-DEM
BEGIN
    SET Raster grid size, sizeGrid //e.g. 1 m, 5 m, 10 m, or 20 m grid resolutions
    SET Projection plain, projPlain = z
    PROJECT (Building-DEM AND Terrain-DEM AND
    Lowlevel-DEM AND Opening-DEM) along projPlain
    WHILE projPlain = z;
        COUNT (rows, columns) along projPlain
        FOR each (row, column) do
            IF active cell at (row, column) contain terrainGrid AND
            wallGrid
                ASSIGN-GRID as wallGrid with sizeGrid-criteria
            ELSE IF active cell at (row, column) contain terrainGrid AND
            lowlevelGrid
                ASSIGN-GRID as lowlevelGrid with sizeGrid-criteria
            ELSE IF active cell at (row, column) contain terrainGrid AND
            (lowlevelGrid AND wallGrid)
                ASSIGN-GRID as wallGrid with sizeGrid-criteria
            ELSE IF active cell at (row, column) contain terrainGrid AND
            (lowlevelGrid AND openingGrid)
                ASSIGN-GRID as lowlevelGrid with sizeGrid-criteria
            …
            ELSE
                ASSIGN-GRID as nullGrid with sizeGrid-criteria
            END IF
        END FOR
        FOR each (row, column) do
            CREATE-DEM use (lowlevelGrid, buildingGrid, wallGrid,
            terrainGrid, nullGrid) with sizeGrid-criteria, MSV-DEM
        END FOR
    END WHILE
END

Table 5-1 Choices of key features for multi-source views (MSV) grid assignments

	'null'	'wall'	'lowlevel' & 'wall'	'lowlevel' & 'opening'	'lowlevel'
'null'	'null'	'wall'	'wall'	'lowlevel'	'lowlevel'
'terrain'	'terrain'	'wall'	'wall'	'lowlevel'	'lowlevel'
'building'	'building'	'building'	'building'	'lowlevel'	'building

In raster-based topographic data, higher resolution results should enable a better description of urban features. A coarse raster grid may misrepresent key components of local features. In contrast, when a sufficient density of raw point cloud data is available, local features can be observed in better detail (Fig. 5-13).

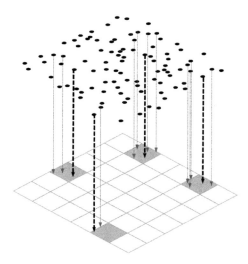

Fig. 5-13. An example of rasterizing point cloud data for
creating raster-based topographic data

When the grid size becomes coarse, the detailed descriptions can disappear. Many examples show that more details of urban feature descriptions can be represented much clearer in high resolution and the coarse resolutions may not adequately represent the local characteristic of urban areas. However, trimming the top section of such multi-source views point cloud data can maintain such pathways, which can also be clearly shown in different spatial resolutions (Fig. 5-14) at 1 m, 5 m, and 10 m, resp.

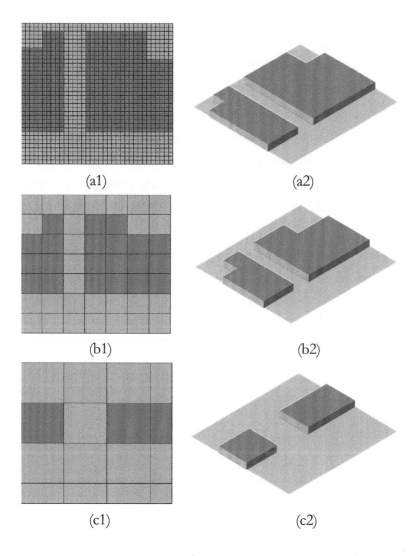

Fig. 5-14. Three examples of MSV-DEM in different grid resolutions: at 1 m grid (a1) plan view (a2) perspective view; 5 m grid (b1) plan view (b2) perspective view; and 10 m grid (c1) plan view (c2) perspective view

5.2.1 Different stages of 2D dynamic flow modelling

During the last decades, advanced improvements in 2D urban-flood modelling offer potentials to predict local flood patterns (e.g. flood inundation areas and flood depths) closer to reality, also simulate flood flow dynamics (e.g. flood velocities and flood flow routings) more accurately at much more efficient computational cost. Efficiency and accuracy are two major indicators for evaluating 2D urban-flood model performances. Typically, efficiency and accuracy are commonly in conflict with one to another (Chen et al., 2012a). While accuracies of adopting finer scales of spatial and temporal resolutions and enhancing 2D model schematics with more key components of local urban features should improve higher accuracy for 2D model simulation, efficiencies of reducing resolutions in space and time scales should also be taken into account for simplifying complexities, speeding up processes, and minimising computational cost. The scale factors of developing such appropriate qualities for 2D model schematics are crucially depending on factors of time scales and space scales, which could be different for flood simulation applications and purposes for each situation. In some situations, urban floods may rapidly rise (the 'dynamic' part in Fig. 5-15) at the initial and thereafter hardly change for the rest of several days or weeks (the 'static' part in Fig. 5-15).

Fig. 5-15. An example of inundation extent changes of (left) dynamic part and (right) static or even stagnant part

In the initial stages of flood events, rapid changes occur (the 'dynamic' part in Fig. 5-15). Increasing a finer time step may capture more significant details of flow dynamic changes for such situations. Moreover, details of obstacles may change flood velocities and they may convey flood flow routes differently. Higher spatial resolutions with key components of complex urban structures could also play a significant role for replicating complex flood dynamics. Floodwater depths and flood inundation patterns may slowly change after some time. Increasing a finer time step may not give significant benefits of representing flow dynamic changes, since the flood is almost static (stagnant) Adopting a simple flat terrain with lower resolutions should be adequate to represent flood inundation maps. Such maps provide information for analysing flood depth damages.

5.2.2 Equivalent roughness

In order to compensate for coarse grid representation of fine grid openings, the roughness values have to be adjusted. Néelz & Pender (2007) adjusted roughness values of coarse grids to account for obstructions. Typically, there is less flow through small openings compare with large openings. In order to compensate for that in a coarse grind model, a simple equivalent roughness relation has been developed for adjusting local roughness of opening pathways in 2D model schematisation as given in Eq. 5-1.

$$M_{coarse} = \frac{O_{coarse}}{O_{fine}} \; x \; M_{fine} \qquad \text{Eq. 5-1}$$

where: M_{coarse} is the new adjusted Manning's M roughness coefficient ($m^{1/3}$ s^{-1})

M_{fine} is the current Manning's M roughness coefficient ($m^{1/3}$ s^{-1})

O_{coarse} is the opening width of the 2D model schematics (m)

O_{fine} is the opening width from the measurements or from the fine topographic data (m)

Adopting such roughness adjustments should be adequate for representing flood dynamic changes, especially at the initial evolution of the flood event. However, when floods have reached their static stage, flow velocities – and resistances – have become much smaller and may not show significant changes.

5.2.3 Urban inundation mapping

In very slow flow – or even stagnant – situations flood velocities and flood flow routings may no longer play an important role. In this case, changes of inundation areas are relatively marginal and incorporating complex urban features into a 2D model schematisation may not be relevant. Applying only terrain LiDAR data (DTM) without complex urban features should be adequate for replicating flood depths and related flood depth damages.

In flat terrain LiDAR data (i.e. LiDAR-DTM), obstacles can be removed by using filtering algorithms (Section 3.4). As a result, LiDAR-DTM contains only ground elevation values see location 'A' in Fig. 5-16a compared to Fig. 5-16b. Either building rooftops or high trees no longer remain. When using the LiDAR-DSM as input for 2D flood simulation, the simulated floodwaters tend to get trapped in local terrain depressions. Buildings and walls can act as impermeable obstacles. Filtering algorithms applied for creating LiDAR-DTM can avoid these apparent barrier of high obstacles.

Simulation results show floodwaters can now freely flow and tend to show more flood inundated areas (Fig. 5-17b). It is clear that in this case floodwater flow directions could not be correctly represented. In this research, exploring the benefits of adopting the new MSV topographic data for enhancing 2D urban flood simulation are carried out for real complex flood situations in complex cities for two case studies: (i) the case study of Kuala Lumpur in Chapter 6, and (ii) the case study of Ayutthaya in Chapter 7.

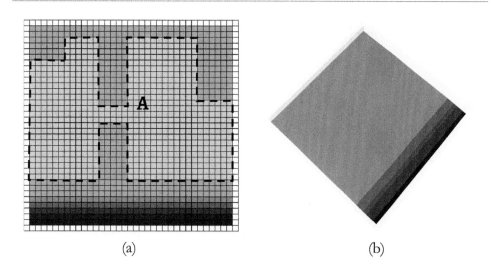

(a)

(b)

Fig. 5-16. Examples of 1 m grid resolution LiDAR-DTM represented in:
(a) a plan view and (b) a perspective view

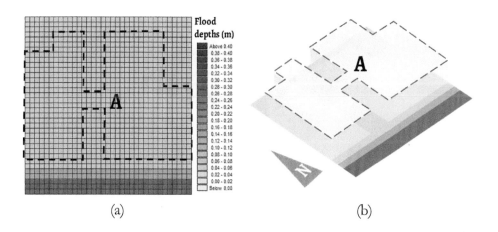

(a)

(b)

Fig. 5-17. Examples of flood simulation using LiDAR-DTM as input represented in:
(a) a plan view flood-depth map with scales on the right and
(a) a perspective view flood-depth map

5.3 Considerations for raster-based topographic data

In a large scale urban scene, applying conventional top-view LiDAR point cloud data could also misrepresent some key features (i.e. low-level structures, walls, and openings) and therefore mistranslate such key features in the conventional LiDAR-DEMs, e.g. LiDAR-DSM (Fig. 5-18a) and LiDAR-DTM (Fig. 5-18b).

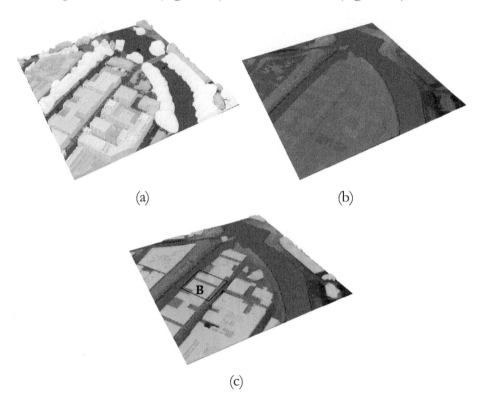

(a)   (b)

(c)

Fig. 5-18. Three examples of raster-based topographic data (a) a LiDAR-DSM,
(b) a LiDAR-DTM, and (c) a new MSV-DEM
(area 'B' in Fig. 5-14)

By merging top-view LiDAR data and side-view SfM data, one could reduce such missing key features and create the novel Multi-Source Views Digital Elevation Model (MSV-DEM) to be used as topographic input data for complex 2D model schematisation (Fig. 5-18c). In the raster-based MSV-DEM, higher resolution results will enable a better description of physical characteristics of complex cities. For coarser grid sizes, detailed features will be smeared or degraded (Fig. 5-19).

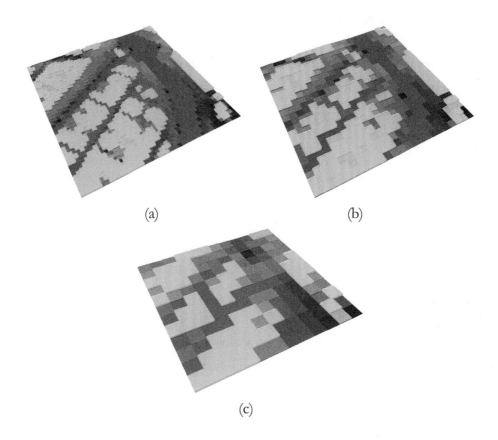

(a)   (b)

(c)

Fig. 5-19. Three different examples of the new MSV-DEM at
(a) 5 m, (b) 10 m, and (c) 20 m grid resolutions

For the new MSV-DEM, errors in data processing can also accumulate but these new MSV data are better representing specific urban features of complex cities. Finer cell boundaries of buildings could better fit the actual building shapes than coarser cell boundaries (Fewtrell et al., 2008; Wang et al., 2010). Technically, a higher spatial resolution could represent more details of urban features than for coarser spatial resolution. However, they may not always provide better flood simulation results in case some key components are not represented appropriately.

5.4 Selection of case study areas

5.4.1 Criteria for selection of case study areas

The primary is focus of this research is on simulation of flood dynamics in and around complex urban features. The study areas were carefully chosen following the criteria below:

1. *Flood prone areas.*
2. *Presence of complex urban features (urban fabrics).*
3. *Availability of topographic data (i.e. top-view LiDAR data).*
4. *Availability of initial and boundary input data for urban flood models.*
5. *Availability of measured data for urban flood model calibrations/verifications.*

5.4.2 Case study area descriptions

The two case studies of Kuala Lumpur, Malaysia and Ayutthaya Island, Thailand were chosen according to the criteria mentioned in Section 5.6.1. The first case study area is Kuala Lumpur, the capital city of Malaysia, which is a densely populated urban area. With an estimated population of over 3.6 million, this city is growing at almost 5% a year. The city is situated in the flood-prone area located at the confluence of Gombak and Klang Rivers. Similar to other flood-prone cities, river banks are often crowded with buildings and constructions with the rapid urbanisation. Kuala Lumpur City still experiences considerable flooding from Klang River and Gombak River. The 2003 flood event of Kuala Lumpur, Malaysia is studied in Chapter 6. An example of overarching urban features hiding their pathways underneath is given in Fig. 5-20.

Fig. 5-20 An example of arch and its pathway underneath

Sky train tracks or elevated highways not only hide underneath pathways but also misrepresent some small urban features, e.g. the kerb underneath (Fig. 5-21).

Fig. 5-21 An example of sky-train track and its low-level structures
underneath, i.e. roads and kerbs

The second case study is Ayutthaya Island, which is also a flood-prone area located in the middle of Chao Phraya Floodplain. In 1991, Ayutthaya Historical Park was declared a UNESCO World Heritage Site. Ayutthaya Island was the former capital city of Siam Kingdom and it is surrounded by rivers. Ayutthaya City is located in the middle of the island and surrounded by three main rivers: Chao Phraya River, Pasak River, and Lopburi River with one Muang Canal. Nowadays, Ayutthaya City are rapidly growing. The city still suffers from frequent floods, especially river flooding. The 2011 flood event of Ayutthaya Island, Thailand is studied in Chapter 7, including extracting watermarks (Fig. 5-22).

Fig. 5-22 An example of flood watermark on concrete wall

Table 5-2 lists key features that are addressed in the two case studies (Kuala Lumpur and Ayutthaya).

Table 5-2 Extracted urban features using MSV data on two case studies:
(i) Kuala Lumpur, Malaysia and (ii) Ayutthaya, Thailand

Urban features	Kuala Lumpur, Malaysia	Ayutthaya, Thailand
(i) Arches	✓	✗
(ii) Openings	✗	✓
(iii) Sky-train tracks	✓	✗
(iv) High trees	✓	✓
(v) Alleys	✓	✓
(vi) Low-level structures	✓	✓
(vii) Watermarks	✗	✓

CHAPTER 6
Applying multi-source views DEM to the case study of Kuala Lumpur, Malaysia

The conventional top-view LiDAR data have some difficulties in representing vertical structures (e.g. building walls, alleys, and openings), and low-level structures (e.g. pathways and kerbs). Such structures are commonly be hidden underneath other elevated pathways, overarching structures, and/or trees. These missing urban features may play a significant role in flow dynamics of complex urban flooding. When applying conventional top-view LiDAR data as input for urban flood models, the simulated results seem to represent mismatches in both floodwater depths and flood propagation patterns. On the contrary, when employing the novel multi-source views (MSV) DEM as input, the simulated results represented good agreements with six measured floodwater depths. These simulated results can also expose and delineate some missing flood extents better than adopting such conventional top-view data as input. In this chapter, a complex urban area of Kuala Lumpur, Malaysia is chosen as a case study (Section 6.1) to simulate the extreme the 2013 flood event. Three different raster-based topographic data: (i) a top-view LiDAR-DSM, (ii) a top-view filtered LiDAR-DTM, and (iii) a novel MSV-DEM, are created (Section 6.2) and then use as input for a coupled 1D-2D numerical model (Section 6.3). The quantitative assessments of the simulated results are represented in Section 6.4. Analysing these results in degrees of flood hazard characteristics are discussed (Section 6.5). Findings are also concluded in Section 6.6.

6.1 The case study

Kuala Lumpur, the capital city of Malaysia, is a densely populated urban area. With an estimated population of over 3.6 million, this city is growing at almost 5% a year. The city is situated in the flood-prone area, located at the confluence of Gombak and Klang Rivers. Almost the same as others flood-prone cities, along the river banks are crowded with buildings and constructions with the rapid urbanisation. Even though major flood mitigation works of Kuala Lumpur City have been implemented already, the city still experiences considerable flooding from rivers (Klang and Gombak Rivers). Yet, the intense tropical rainfalls occasionally overwhelm the designed drainage capacities. Damages and losses cannot be completely avoided, but flood preparedness can considerably improve, at least, by enhancing 2D urban flood simulation.

6.1.1 Description of the case study

A small area of Kuala Lumpur was chosen as a case study for exploring effects of using different raster-based topographic data as input for 2D urban flood simulation. Some data measured during the major flood event were available and a 1D numerical flood model had already been developed by DHI Water & Environment (DHI, 2004). The complex urban area covered about 0.48 km², located at the confluence of the Klang River flowing from the Northeast, and the Gombak River flowing from the Northwest (Fig. 6-1).

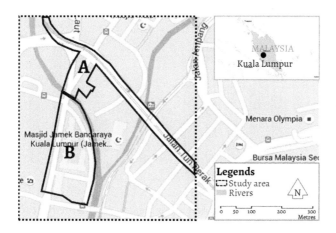

Fig. 6-1.   A study area of Kuala Lumpur City (background: Google Maps™, 2012a). The small map shows the location of the city (top right corner) with the legends (low right corner). Two locations 'A' and 'B' were explored by side-view SfM surveying

6.1.2 Climate and rainfall patterns

In this region, the weather is generally hot and wet through the year. The Kuala Lumpur's climate is characterised by uniform high temperature with high relative humidity, heavy rainfall and little wind. The temperature is relatively constant of ~27°C. The lowest and highest temperatures are between 23°C to 32°C with the relative humidity of ~82%.

The average annual rainfall depth in this region is about 2,400 mm, which can be defined by the following seasons: (i) the northeast monsoon started from December to March, (ii) a transitional period from April to May, (iii) the southwest monsoon from June to September, and (iv) a transitional period from October to November. The monthly variation of the rainfall was given (Fig. 6-2).

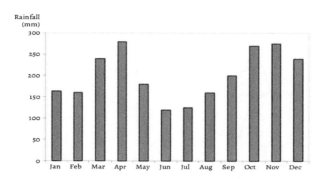

Fig. 6-2.   Average monthly rainfall in Kuala Lumpur recorded from the year 1996 to 2003,
source: Malaysian Meteorological Department (MMD)

The highest rainfall of ~280 mm occurs in the months of April, October, and November. The lowest rainfall of ~115 mm takes place in June. The rainy seasons occur in the transitional periods between March and April and from October to November. Typically, very high-intensity rainstorms occasionally occur and last for a short period.

6.2 Topographic data acquisition and rasterization

The three different types of the raster-based topographic data were constructed. Top-view LiDAR data are provided by Drainage and Irrigation Department (DID), Malaysia. The former two topographic data were created by using a single top-view approach. The raw LiDAR data was used to create (i) a top-view LiDAR digital surface model (LiDAR-DSM) and (ii) a top-view filtered LiDAR digital terrain model (LiDAR-DTM), elaborated in Section 6.2.1 and 6.2.2 resp. A so-called multi-source view (MSV) approach had been used to create the latter topographic data. By merging a top-view filtered LiDAR data and a side-view structure-from-motion (SfM) data (Section 6.2.3), the new multi-source views digital elevation model (MSV-DEM) was created, as described in Section 6.2.4.

6.2.1 7Top-view LiDAR digital surface model (LiDAR-DSM)

The LMS-Q560 LiDAR system with 75 kHz of effective signal and 60° field of view (FoV) was set and mounted on a Bell-206b Jet Ranger helicopter. This helicopter flew at an altitude of ~700 m with an averaged ground speed of 51.4 m s⁻¹ to maintain 40% side lap at each flying path. An inertial measurement unit (IMU) was used, and a global positioning system (GPS) provided absolute locations and elevations (referencing the EPSG 32647: WGS84 UTM 47N coordinate system) with the accuracy of ±0.05 m in the horizontal direction and approximately twice as much in the vertical direction. The LiDAR data were collected with an average single run density of ~2.4 points per meter or a diameter of ~42 cm between each elevation point. Therefore, urban features smaller than this diameter could not be captured in this dataset.

By using a traditional single top-view approach, the structures, which mistakenly perceived as obstructions in the rivers had been removed manually. The neighbouring elevations of river profile were then interpolated to fill in the gaps. The isolated points that were less than ten neighbouring points within 1 m horizontal distance were thinned and simplified to create top-view LiDAR point cloud data. These point cloud data were rasterized at 1 m grid resolution for creating a top-view LiDAR digital surface model (LiDAR-DSM) using ArcGIS®. This LiDAR-DSM (Fig. 6-4a) was then translated to the same grid resolution for 2D model schematics using Grd2Mike tool in Mike Zero™. Details of top-view LiDAR data extraction and rasterization processes were given in Section 3.4.

6.2.2 Top-view filtered LiDAR digital terrain model (LiDAR-DTM)

Following from the previous step, the filtered point cloud of top-view LiDAR data was constructed. The sky train tracks, high trees, and elevated road (Fig. 6-3a) that may be perceived as floodwater obstructions have been removed by using filtering algorithms followed Abdullah et al. (2012b). After that, these obstacles were then removed from point cloud data. The neighbouring elevations were then interpolated to fill in the gaps.

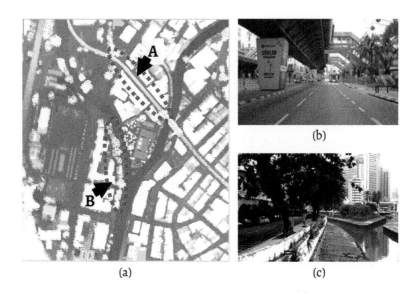

Fig. 6-3.   An example of (a) A LiDAR-DSM with river crossing structures: (b) a sky-train track and bridge underneath (location 'A' in Fig. 6-3a); (c) high trees and kerbs underneath (location 'B' in Fig. 6-3a)

The filtered point cloud of top-view LiDAR data was rasterized at 1 m grid resolution for creating a top-view LiDAR digital terrain model (LiDAR-DTM) using ArcGIS®. This LiDAR-DTM (Fig. 6-4b) was then translated to the same grid resolution for 2D model schematics using the Grd2Mike tool in Mike Zero™. Details of top-view LiDAR data extraction and rasterization processes were given in Section 3.4.

(a)   (b)

Fig. 6-4.   (a) A top-view LiDAR digital surface model (LiDAR-DSM).
(b) A top-view filtered LiDAR digital terrain model (LiDAR-DTM)

6.2.3 Side-view SfM surveying

To obtain topographic data from side viewpoints, a Nikon D5100 digital single-lens reflex (DSLR) camera was used as a video camcorder to record high definition videos of 2 M pixels at 30 frames per second. More than 12 video scenes were recorded from side-view along the streets of Jalan Tun Perak Road, and of the Bangunan Sultan Abdul Samad Building known as the high court of Kuala Lumpur City (locations 'A' and 'B' in Fig. 6-1). By using Python scripts, an automated image separation slices the video scenes into a series of overlapping photos with the same resolution.

In this section, some examples of SfM point cloud illustrate the significantly urban features (Fig. 6-5a): a sky-train track with kerbs underneath (Fig. 6-5b1 and b2); high trees with retention wall underneath (Fig. 6-5c1 and c2); arch building with pathway underneath (Fig. 6-5d1 and d2).

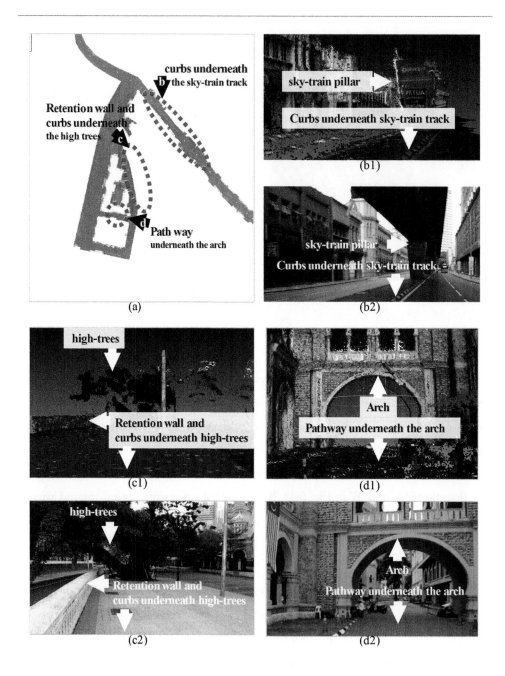

Fig. 6-5. (a) Areas of side-view SfM exploration: the sky-train track and its kerbs underneath shown in (b1) a side-view SfM point cloud image and (b2) a photo; high trees with retention wall and kerbs underneath shown in (c1) a side-view SfM point cloud image and (c2) a photo; the arch and its pathway underneath showed in (d1) a side-view SfM point cloud image and (d2) a photo

Even though the vibration reduction (VR) technology of the Nikkor 18-55 mm lens was set to stabilise the captured video scenes, some blurry scenes due to camera motion or object movement were still recorded. A subsequent removal of blurred photos was undertaken manually. A laptop, running 64-bit Microsoft Windows 7 equipped with four physical cores Intel® i7 CPU at 2.20 GHz, 16 GB of RAM, and 2 GB Video RAM embedded in NVIDIA® GeForce™ GTX580M graphics cards was used for point cloud data processing. Twenty distinct ground control points (GCPs) in the LiDAR dataset were used as geo-referencing positions to the SfM point cloud data. The absolute error in producing side-view SfM data found to be in the order of ~18 cm (RMSE in the GCP data). The further steps of the raw SfM data processing and registration were explained in Section 4.4. Side-view SfM key component extraction was also given in Section 4.5.

6.2.4 Multi-source views of digital elevation model (MSV-DEM)

As described above, the single-view approach was applied to create both top-view LiDAR point cloud data and side-view SfM point cloud, focusing on the sub-meter spatial resolution. By fusing of 3D point cloud obtained from different data sources and different viewpoints, a novel technique based on multi-source views approach was introduced.

First, top-view LiDAR data (a square dotted area in Fig. 6-1) and side-view SfM data (locations 'A' and 'B' in Fig. 6-1) were manually fused to create the multi-source views (MSV) point cloud data by using MeshLab® open source software. Some example of MSV point cloud data are demonstrated, e.g. the fused data obtained from top-view LiDAR data with side-view SfM data (Fig. 6-6). Consequently, the MSV data can also have benefited from both a large coverage area of top-view LiDAR data and a great level of detail of side-view SfM data (~18 cm resolution).

Fig. 6-6. An example of the multi-source views point cloud obtained by
merging top-view LiDAR data with side-view SfM data

However, when employing such complex and high-resolution point cloud data are
still too complex for translating to a simple 2D model schematic. Top-view LiDAR
point cloud and side-view SfM point cloud were simplified, separately. Scale factors
of creating grid-based resolutions for 2D schematics should adequately represent
all essences of key features (i.e. low-level structures, walls, and openings) and
therefore translate them in their coarse or fine-resolution model schematic cells.

In this section, top-view LiDAR point cloud and side-view SfM point cloud were
extracted and simplified in ArcGIS® by thinning the isolated points, which were
less than ten neighbouring points within a 1 m horizontal distance. Merging of
these different-source views DEMs were then carried out in order to create new
multi-source views digital elevation model (MSV-DEM) using ArcGIS®. This
MSV-DEM (Fig. 6-7) was then translated to the same 1 m grid resolution for 2D
model schematics using Grd2Mike tool in Mike Zero™. Details of merging multi-
source view data were given in Section 5.3.

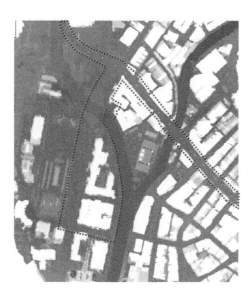

Fig. 6-7.   The multi-source views of digital elevation model (MSV-DEM) merging top-view
filtered LiDAR data with side-view SfM data. The dotted line refers to
the boundaries of a side-view SfM surveys

6.3   Numerical modelling schemes

A coupled 1D-2D model of Kuala Lumpur had been further developed to investigate the propagation of excess floodwater from the 1D river system of the Klang and Gombak Rivers into the 2D urban area (Fig. 6-8), using the Mike FLOOD™ software by DHI™. Three different raster-based topographic data were generated and used as input for a 2D urban flood modelling, as described hereafter. A Manning's M friction coefficient of 40 was applied uniformly to the constructed 1D river networks, following the criteria defined by Chow (1959). The Manning's M of 30 was used for the 2D urban surface area, applied identically to each of the three raster-based topographic data following an earlier study by Abdullah et al. (2012a); Abdullah et al. (2012b).

The flood simulation results were evaluated and compared with the six measurement locations (six measurement locations in Fig. 6-10) observed by the Department of Irrigation and Drainage (DID).

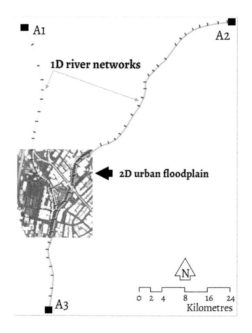

Fig. 6-8.   A schematic of the coupled 1D-2D modelling
(1D river network and 2D floodplain area)

The 2003 flood event that occurred on June 10th, 2003 was used as the case study, observed by the Drainage and Irrigation Department (DID). At the Gombak River, the discharges at Jalan Tun Razak station (location 'A1' in Fig. 6-8) were over 149 m³ s⁻¹ (Fig. 6-9a). At the Klang River, the water levels at Tun Razak Bonos station (location 'A2' in Fig. 6-8) were over 34 m msl (Fig. 6-9b), and the water levels at Jambatan Sulaiman station (location 'A3' in Fig. 6-8) were above 28 m msl (Fig. 6-9c). Further to that, the water levels in rivers started to spill over the banks onto the Kuala Lumpur City.

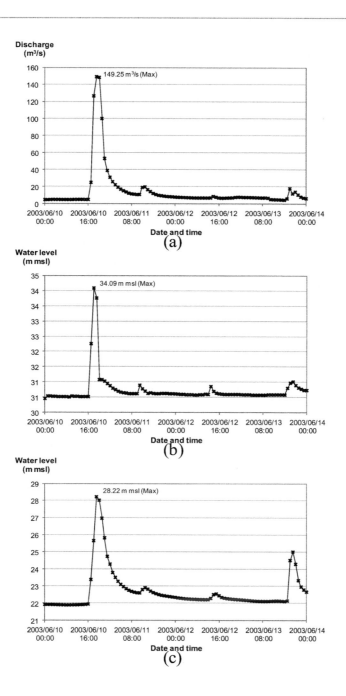

Fig. 6-9. Time-series data of discharges at (a) Jalan Tun Razak and (b) water level at Tun Razak Bonos. Time-series data of water levels at Jambatan Sulaiman

6.4 Results

Flood simulation results have been evaluated of six locations (marked "1" to "6" in Fig. 6-10). We can observe that the largest difference (58%) in floodwater depths occurred at Jalan Parlimen (Table 6-1; location '4' in Fig. 6-10).

Table 6-1 Comparison of numerical simulations versus measured floodwater depths at six locations for three different input raster-based topographic data: (%Diff from Eq. 3-1)

| | Dataran Merdeka(1) % | | Leboh Ampang(2) % | | Jalan Meleka(3) % | |
	Flood depth (m)	Diff.	Flood depth (m)	Diff.	Flood depth (m)	Diff.
1D-2D model using:						
Measurement	0.50		1.20		1.30	
LiDAR-DSM	0.68	36.0	1.39	15.8	1.49	14.6
LiDAR-DTM	0.46	8.0	1.29	7.5	1.43	10.0
MSV-DEM	0.47	6.0	1.28	6.7	1.40	7.7

| | Jalan Parlimen(4) % | | Jalan Raja(5) % | | Leboh Pasar(6) % | |
	Flood depth (m)	Diff.	Flood depth (m)	Diff.	Flood depth (m)	Diff.
1D-2D model using:						
Measurement	0.50		1.00		0.65	
LiDAR-DSM	0.79	58.0	1.22	22.0	0.85	30.8
LiDAR-DTM	0.46	8.0	1.08	8.0	0.72	10.8
MSV-DEM	0.47	6.0	1.04	4.0	0.68	4.6

When applying the MSV-DEM, the simulation results show that fewer obstructions could have larger flood propagation areas (Table 6-2).

Table 6-2 The computed inundation areas of 1D-2D simulations
using three different raster-based topographic data as input

	Inundation extent (m²)
1D-2D model using:	
LiDAR-DSM	128,870
LiDAR-DTM	156,824
MSV-DEM	151,699

When applying a LiDAR-DSM as topographic input, the maximum depth in the simulation differs from the measurements by about +0.29 m. Conversely, when applying a LiDAR-DTM and MSV-DEM as input, these values are seen to reduce considerably down to 8.0% and 6.0%, resp.

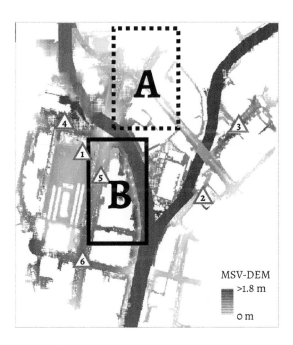

Fig. 6-10. Maximum flood depths using MSV-DEM as input; marked locations 1 to 6 indicate the maximum floodwater depth observed by DID

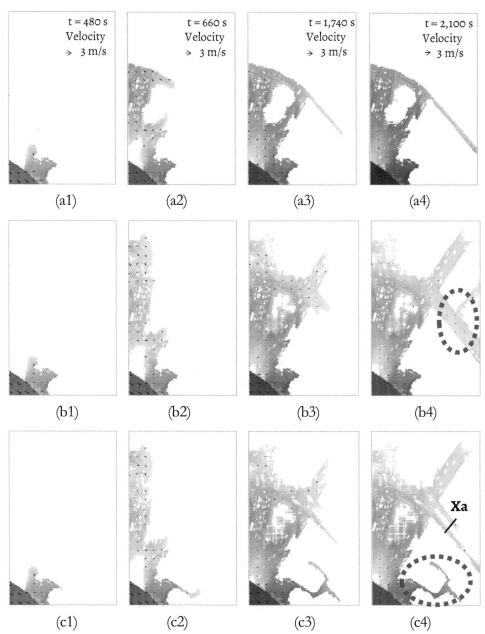

Fig. 6-11. Flood simulation results (floodwater depths and flow velocities) in 'A' sub-region (location 'A' in Fig. 6-10) at time steps t = 480 s, 660 s, 1,740 s, and 2,100 s using: (a1 to a4) LiDAR-DSM, (b1 to b4) LiDAR-DTM, (c1 to c4) MSV-DEM, as topographic input data. The cross section 'Xa' in (c4) runs SW-NE, referred to in detail in Fig. 6-13

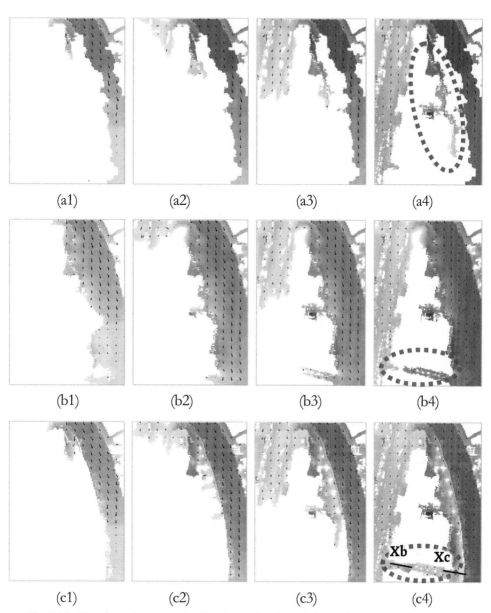

(a1)   (a2)   (a3)   (a4)

(b1)   (b2)   (b3)   (b4)

(c1)   (c2)   (c3)   (c4)

Fig. 6-12. Flood simulation results (floodwater depths and flow velocities) in 'B' sub-region (location 'B' in Fig. 6-10) at time steps t = 480 s, 600 s, 750 s, and 1,140 s using: (a1 to a4) LiDAR-DSM, (b1 to b4) LiDAR-DTM, (c1 to c4) MSV-DEM, as topographic input data. Cross sections 'Xb' and 'Xc' in (c4) run W-E, elaborated in Fig. 6-14 and Fig. 6-15, resp

From Fig. 6-10, it shows that the maximum flood depth was better simulated when using new MSV-DEM as topographic input. From the 'A' and 'B' sub-regions in Fig. 6-10. The detailed results of the time evolution of floodwater depths and flow velocities showed that the floodwaters primarily propagate along the riverbanks, roads, pathways, and along with other lowland areas ('A' and 'B' sub-region results in Fig. 6-11 and Fig. 6-12, resp). In 'A' sub-region, the flow was from the South to North. Floodwaters start to overflow from the Gombak River into the city at time step t = 480 (Fig. 6-11, b1, and c1). In 'B' sub-region, floodwaters start to overflow from the Gombak River to the riverbank at the same time step t = 480 s (Fig. 6-12, b1, and c1).

6.4.1 Simulated results using the LiDAR-DSM

When applying a LiDAR-DSM as input data, the flood simulation results in 'A' sub-region show that higher flood depths accumulate in the middle part at time step t = 660 s and the flows are seemingly blocked by the sky train track, diverting the flows towards the S-E direction (Fig. 6-11a1 to a4). In 'B' sub-region, the floodwaters flow in the N-S direction (Fig. 6-12a1 to a4). The flood simulation results show that floodwaters accumulate along the riverbank in the Eastern part, and these floodwaters are confined between high trees and buildings (a circular dotted area in Fig. 6-12a4).

6.4.2 Simulated results using the LiDAR-DTM

When applying a LiDAR-DTM as input data, it showed that the floodwaters in 'A' sub-region can freely flow towards the Northern part of the domain without being blocked by the sky-train track (Fig. 6-11b1 to b4). In 'B' sub-region, the floodwaters flow in the N-S direction (Fig. 6-12b1 to b4). Floodwaters start to overflow directly from the Gombak River in the E-W part of the domain. The floodwaters flow over the riverbank towards the pathways before they are seemingly blocked by the arch of the Bangunan Sultan Abdul Samad Building at t = 750 s. It appears as if the floodwaters coming from the East cannot reach the floodwater from the West because of the arch (a circular dotted area in Fig. 6-12).

6.4.3 Simulated results using the new MSV-DEM

Even though applying either the LiDAR-DTM or MSV-DEM shows that floodwaters can now freely flow into the Northern part of the domain in 'A' sub-region, the simulation results still illustrate that some inundated parts could only be exposed when employing the MSV-DEM (a circular dotted area in Fig. 6-11) which was capable of capturing small urban features (e.g. pathways and kerbs). In 'B' sub-region, floodwaters do not come directly from the river in the East, but first start to overflow from the Gombak River in the North before flowing into the urban area in the South. Small urban features such as retention walls and kerbs will divert some floodwaters flow along riverbanks and alleyways (Fig. 6-12c1 to c4). In addition, floodwaters can connect with the other side at t = 1140 s (a circular dotted area in Fig. 6-12).

When considering the longitudinal profile of 'Xa' cross section (Fig. 6-11), it can be observed that for LiDAR-DSM as input. The floodwaters are confined to the zone on the left side (Fig. 6-13a), while applying LiDAR-DTM shows that the floodwaters inundate most of the cross section (Fig. 6-13b). On the other hand, using MSV-DEM, flood waters are diverted and separated by small kerbs (Fig. 6-13c).

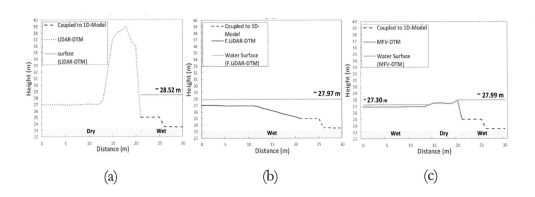

Fig. 6-13. Simulated water surface and elevation profiles for (a) LiDAR-DSM, (b) LiDAR-DTM, and (c) MSV-DEM, at time steps t = 2,100 s, at 'Xa' cross section

The second longitudinal profile of 'Xb' cross section (Fig. 6-12) shows floodwater depth patterns in more detail. When applying LiDAR-DSM as input, floodwaters are confined to the Gombak River on the left side (Fig. 6-14a). Applying LiDAR-DTM shows that floodwaters are seemingly blocked by the arch of the Bangunan Sultan Abdul Samad Building (Fig. 6-14b), showing the separate wet zone on the left side (riverside) and another on the right side (the roadside). On the other hand, using MSV-DEM, flood waters can flow freely from the riverside into the roadside, or the other way around (Fig. 6-14c).

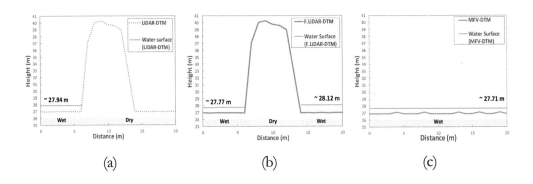

Fig. 6-14. Simulated water surface and elevation profiles for (a) LiDAR-DSM, (b) LiDAR-DTM, and (c) MSV-DEM, at time steps t = 1,140 s, at 'Xb' cross section

The third longitudinal profile of 'Xc' cross section (Fig. 6-12) also shows that when applying LiDAR-DSM as input, floodwaters are confined to the wet zone of the Gombak River on the right side (Fig. 6-15a). Applying LiDAR-DTM shows that floodwaters flow freely from the river on the left side to the riverbank on the right side (Fig. 6-15b). Using MSV-DEM, floodwaters are (seen to be) diverted by small retention walls and kerbs along the riverbank, separating the floodwaters from the left (riverbank) side to the right (river) side (Fig. 6-15c).

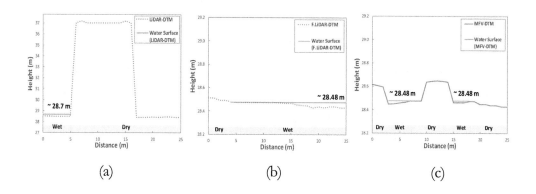

Fig. 6-15. Simulated water surface and elevation profiles for (a) LiDAR-DSM, (b) LiDAR-DTM, and (c) MSV-DEM, at time steps t = 1,140 s, at 'Xc' cross section

6.5 Discussion

The flood simulation results presented here show that incorporation of urban features (sky-train tracks, elevated roads, high trees) in a numerical model is an important aspect as such features can cause and can play a significant role in diverting the shallow flows that are generated in urban environments.

When using (i) a LiDAR-DSM input (Fig. 6-4a) as topographic input data, some high features (i.e. sky-train tracks) are seen to behave as dykes (Fig. 6-13; location 'Xa' in Fig. 6-11c4). The floodwaters seem to propagate only in the Southern direction and no floodwater appears in the Northern part of the domain (Fig. 6-11a2 to a4; 'A' sub-region location in Fig. 6-10). In 'B' sub-region, high trees (the circular dotted area in Fig. 6-12a4) look like an island (Fig. 6-15; location 'Xc' in Fig. 6-12c4), seemingly obstructing floodwater flows from the Gombak River in the East.

When applying (ii) a LiDAR-DTM (Fig. 6-4b) as input data, the filtered data replace high features (sky train tracks and high trees) with ground-flatted elevations. However, it is not possible to correctly create or reconstruct the small urban features hidden underneath high features. Nevertheless, such (hidden) small urban features can have a considerable effect on floodwater dynamics and predictions (Remondino & El-Hakim, 2006; Rychkov et al., 2012; Sampson et al., 2012). By applying the LiDAR-DTM as input, flood simulation results have misrepresented floodwater propagations (a circular dotted area in Fig. 6-11c4 compare to Fig. 6-11b4) and mismatch floodwater depths (Fig. 6-13; location 'Xa' in Fig. 6-11c4).

When including SfM data, it is possible to take into account small urban features by using (iii) the multi-source views (MSV) approach. This new approach can represent small details of urban features that can still play a significant role in diverting shallow floodwaters. As a result, floodwater propagation can be predicted much better. It is observed that small kerbs (sideways and middle parts of Jalan Tun Perak Road) play a significant role in diverting and confining floodwaters flow along the road. In 'B' sub-region (Fig. 6-10), floodwaters can now flow freely through alleyways without any obstruction by arches. The results also show that small urban features located along riverbanks (i.e. retention walls and kerbs) can play a major role in diverting floodwater patterns (Fig. 6-15; location 'Xc' in Fig. 6-11c2). It can be noted that applying new MSV-DEM can represent more details of riverbank structures as well as hidden small urban features such as pathways, retention walls, and sidewalk kerbs.

6.6 Conclusions

The results obtained from the present work show that the proposed technique based on the fusion of LiDAR data and Structure from Motion (SfM) observations can be very beneficial for flood modelling applications. In the case study presented here, a 1D-2D numerical modelling approach was used to simulate the extreme urban flooding event that occurred on June 10th, 2003 in Kula Lumpur (Malaysia). Three different raster-based topographic data were derived from top-view LiDAR data and side-view SfM observations.

From the analysis, it was found that when employing a LiDAR-DSM the flow patterns and water depths may not be correctly represented in the digital terrain map since overarching structures such as a sky train or elevated roads are usually perceived as obstructions for floodwater propagation. Some obstructing features could be removed by applying some filtering algorithms in the LiDAR-DTM. However, in the filtered DTM map the obstructing features can only be replaced with the surrounding ground-flattened elevations that do not contain particular urban features hidden underneath. Present work has shown that a ground-based SfM technique can be effective in detecting small urban features (e.g. arches, retention walls, alleyways, and kerbs). Correspondingly, flood simulation results found to be not only in good agreement with floodwater depth observations but also to represent floodwater dynamics in better details. Such detailed models are of increasing concern to insurance sectors to evaluate the primary drivers of the flood damages and losses (Evans et al., 2008). It should also help to assess road network drainage capability to both the public sector for critical infrastructure planning and to the private sector for business interruption loss estimation (Sampson et al., 2012).

Overall, it can be concluded that new multi-view approach of combining top-view LiDAR data with side-view SfM observations is capable of creating a more accurate digital terrain map, which can be then used as input for numerical urban flood model simulation and produce a more realistic representation of floodwater dynamics and inundation depths.

CHAPTER 7
Extracting inundation patterns from flood watermarks: the case study of Ayutthaya, Thailand

Flood watermarks stipulate peak water depths from a flood event, indicating a magnitude of inundation that took place. Such information is invaluable for instantiation and validation of urban flood models. However, collecting and processing such data from land surveys can be costly and time-consuming. New remote sensing and data processing technologies offer improved opportunities to address these issues. The new Structure from Motion (SfM) technology and its applications are capable of extracting flood watermarks. In this chapter, an area of World Heritage Site of Ayutthaya Island, Thailand, is chosen as a case study (Section 7.1). Top-view LiDAR data are used as conventional topographic input data to create three different types of DEMs (Section 7.2). For the first time, side-view SfM surveys using two mobile units are implemented for achieving side-view SfM topographic data. The first application of side-view SfM data is used to extract historical flood watermarks (Section 7.3), which are then compared with eleven measurement locations. The work undertaken demonstrates significant capability of SfM technology for extraction of flood watermarks (Section 7.4). The second application is to make use of merging side-view SfM data with conventional top-view LiDAR data to create (iv) the novel multi-source views (MSV) DEM, which in turn have improved schematization of 2D flood models (Section 7.5). Four different DEMs are then used as inputs for coupled 1D-2D flood models (Section 7.6). By applying new MSV-DEM, some highlighted results are given in Section 7.7. Findings are finally discussed and concluded in Section 7.8 and 7.9, resp.

7.1 The case study

The case study area used in the present work is located in the Ayutthaya Island (Thailand). In the past in 1463-1666 and 1688-1767, Ayutthaya Island was the former capital city of Siam Kingdom and it is surrounded by rivers. Ayutthaya's ancestors had long been used these rivers for agriculture, daily consumptions, and transportations, also adapted living with floods to their lives. Exuberances of river networks brought a lot of profits from trading and taxing to the city (Phungwong, 2012). Synchronising river networks with floods also had been used for intercepting infantry enemies for several times. Ayutthaya City and their glories were acknowledged as an ancient international port of the Southeast Asia. The remaining pagodas and gigantic monasteries apparently remind how exquisite the city was in the past. In 1991, Ayutthaya Historical Park declared the UNESCO World Heritage Site under criteria III as an excellent witness to the period of development of a true national Thai art.

The Ayutthaya Island is a flood-prone area and it still suffers from frequent floods. Along riverbanks of Ayutthaya are nowadays crowded with buildings and constructions. From time to time, rapid urbanising of Ayutthaya City still suffers from urban floods, especially river flooding. During the 2011 flood event, this area was under floodwater for almost two months resulting in 97 fatalities and significant damages.

7.1.1 Description of the case study

A province of Ayutthaya is ~65 km further up the North from Bangkok – the capital city of Thailand, and ~90 km from the Gulf of Thailand. Ayutthaya region is surrounded by seven neighbouring provinces: Lopburi and Angthong Provinces to the North; Pathum Thani and Nonthaburi Provinces to the South; Saraburi Province to the East; Suphan Buri and Nakhon Pathom Provinces to the West.

Ayutthaya Island is in the heart of the city, which is geographically delineated by 14°22'04" at the North, 14°20'22" at the South, 100°34'54" at the East, and 100°32'35" at the West, covering the area ~8 km². The three rivers that surround the Island are Lopburi River coming from the North, Pasak River coming from the Northeast, Chao Phraya River coming from the Northwest to the South, and one ancient Lopburi River (known as a Muang Canal), which was previously connected to the Chao Phraya River. The case study area used in the present work is depicted in Fig. 7-1.

Fig. 7-1.   Case study area used in the present work – Ayutthaya Island (background: Google Maps™, 2012b) with the location of Ayutthaya Province on top left corner and legends on low left corner

7.1.2 Climate and rainfall patterns

The climate of Ayutthaya consists of three main seasons, i.e. hot, rainy, and cool. The warmest month is in April when a temperature reaches 30.3°C associated with a daily direct sunlight of 8 to 10 h. The rainy season starts from May to October when a monthly precipitation varies between ~125 mm and ~260 mm with humidity above ~70%. The cool season occurs around November until February when the temperature drops and rainfall decrease (Table 7-1).

Table 7-1 Average measurements of Ayutthaya weather stations recorded monthly from the year 2009 to 2012 by Thailand Meteorological Department, Ministry of Information and Communication Technology (TMD/MICT)

Description	Jan	Feb	Mar	Apr	May	Jun	Jul	Aug	Sep	Oct	Nov	Dec
Temperature(°C)	26.7	28.0	30.1	30.3	30.0	29.3	29.0	28.8	28.2	28.3	27.1	25.9
Humidity (%)	61.2	61.1	63.3	68.6	70.3	71.4	72.5	72.4	75.9	74.0	69.2	62.3
Rainfall (mm)	1.5	14.7	38.0	46.5	163.4	123.6	197.8	216.9	257.0	130.2	29.7	7.0

7.1.3 Severe flooding event in 2011

In 2011, a series of monsoons and tropical storms incessantly approached to Thailand, resulting intensive rainfalls occurred across most of the country. Heavy rains started with the arrival of Haima Tropical Storm on June 24th to 26th, following by Nock-Ten Tropical Storm on July 30th to August 3rd, and accelerating rainfall severities to the North, Northeast, and Central. Rainfalls intensity and magnitude (and consequently floods) further increased with Haitang Tropical Storm that occurred on September 28th, Nalgae Tropical Storm on October 5th to 6th, and Twenty-Four Tropical Depression (Fig. 7-2), resp.

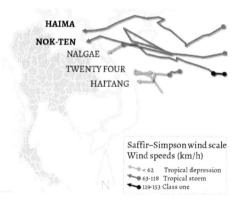

Fig. 7-2.   A series of tropical storms approached to Thailand in 2011 (HAII, 2012)

When the two biggest dams (the Bhumibol Dam and Sirikit Dam) in the North nearly reached their capacity, they started to release excess discharges southward to the Chao Phraya River. The floodwaters then started to propagate over the Chao Phraya floodplain. Moreover, breaching failures of ten major flood-control structures exacerbated the perfect condition to one of the worse flood event in the past 100 years of Thailand (Keerakamolchai, 2014). Thai people were suffering from deaths of 813 fatalities reported in the nationwide. Flood damages were estimated at 46.5 billion USD over the next two years and beyond. Rehabilitation and reconstruction costs were evaluated at 50 billion USD (GFDRR, 2012). Economic growths in the fourth quarter of 2011 were contracted considerably, reducing GDP growth of the country from 2.6% to 1.0% (BOT, 2012). Sixty-six of seventy-six provinces were inundated, of which Ayutthaya had the highest fatality of 97 deaths (BOE, 2011).

In Ayutthaya Island, during the 2011 flooding event, all water gates were closed, and pumps had been intensively operated by Royal Irrigation Department, Ministry of Agriculture and Cooperatives (RID/MOAC). Although Ayutthaya Island resisted to floods for several weeks, eventually, on October 11th, 2011, at 23:00 Indochina Time (ICT), the island was inundated with estimated flood depth of ~2 m and lasted for almost two months (Table 7-2).

Table 7-2 A description of major Ayutthaya inundations over the past 70 years
(source by Toyoda et al., 2012)

Flooding events	Descriptions
1942	Inundation level up to 5.51 m in Ayutthaya and Bangkok for two months.
1983	Multiple inundations in Thailand during five months.
1995	Flooding in some ancient towns, archaeological sites, and monuments in Ayutthaya. Many historic structures and sites were damaged and/or collapsed.
2011	Inundation of ~2 m inside Ayutthaya Island.

Flooding affected several public services, governmental offices, commercial-residential buildings, and historical sites. Damage costs only in the Ayutthaya Historical Park (AHP) were estimated at 2.1 million USD (Toyoda et al., 2012).

7.2 Top-view LiDAR data acquisition and processing

7.2.1 Aerial surveying

The Cessna 404 (OO-MAP) RCD 105 aeroplane was used for aerial based surveys. Different surveying tools (i.e. DGPS, IMU, and LiDAR systems) were installed on a rigid rig in the aeroplane. An IMU was originally used to provide orientations, which can also be used for estimating locations of the aeroplane when the GPS signals were weak. These setups can provide an absolute accuracy of ± 0.03 m in horizontal directions with accuracy twice as much in vertical directions (in post-processing mode). The Leica-ALS60™ (SN6125) LiDAR system with 86.7 kHz of pulse frequency and 60° field of view (FoV) was set to reconstruct topographic point cloud data. With a ground speed of 61.7 m s^{-1}, this aeroplane flew at altitude of 2,780 m above the ground to maintain 30% side lap at each flying path. A beam diameter was 0.63 m, urban features smaller than this diameter cannot be captured into this dataset.

In this case study, raw top-view LiDAR data were distributed by Geo-Informatics and Space Technology Development, Ministry of Science and Technology (GISTDA/MOST) supported by Japan International Cooperation Agency (JICA). These raw LiDAR data were already registered to a georeferencing system of EPSG 32647: WGS84 UTM 47N. The details of top-view LiDAR data acquisition and registration processes were already explained in Section 3.2 and 3.3, resp.

7.2.2 Top-view LiDAR data processing

When registered top-view LiDAR data were ready for further processing steps, some urban structures, which mistakenly perceived as river obstacles (e.g. river crossing structures and vegetation), were manually removed. Removal point cloud data were then replaced by neighbour interpolated elevations. Isolated points that less than ten neighbouring points within 1 m horizontal distance were next thinned and simplified to create top-view LiDAR point cloud using ArcGIS®.

By using building footprint boundaries supported by the Ayutthaya's municipality, building point cloud data were further classified. When building point cloud data were extracted, remaining point cloud data were still combined with point cloud of bare-earth terrains and trees. Then, a morphological filtering, with a referencing elevation of 4 m msl and a threshold of +1 and -2 m, resp. was applied. Elevation points within this threshold can then be classed as a terrain point cloud. This simple and straightforward approach was chosen due to, the study area is relatively flat and small. These terrain point cloud data were next ready to create a raster-based topographic data of (i) top-view LiDAR digital surface models (LiDAR-DSMs; see an example in Fig. 7-3a) and (ii) top-view LiDAR digital terrain models (LiDAR-DTMs; see an example in Fig. 7-3b). A combination of terrain and building point cloud data were used to create (iii) top-view LiDAR digital building and terrain models (LiDAR-DBMs+; see an example in Fig. 7-3c). These extracted urban features (i.e. buildings and terrains) of top-view LiDAR point cloud data were also rasterized at 5 m, 10 m, and 20 m grid resolution for creating top-view LiDAR-DEMs (i.e. Building-DEM and Terrain-DEM) using ArcGIS®. Only three main raster-based LiDAR data, i.e. (i) LiDAR-DSM, (ii) LiDAR-DTM, and (iii) LiDAR-DBM+ were then translated to the same grid resolution for 2D model schematics using Grd2Mike tool in Mike Zero™. Details of top-view LiDAR data extraction and rasterization processes were given in Section 3.4.

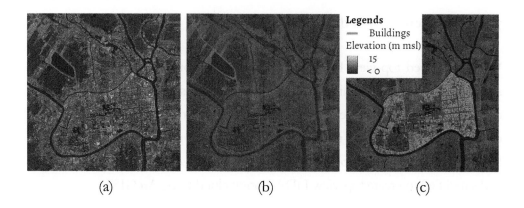

Fig. 7-3. Three examples of different raster-based topographic data at 5 m grid resolution: (a) LiDAR-DSM, (b) LiDAR-DTM, and (b) LiDAR-DBM+ with its legends on top left corner

7.3 Side-view data acquisition and processing

7.3.1 Side-view surveying

Even though several photos can capture and represent some flood levels, most captured photos still have difficulties for representing actual peaks of the flood event. Unlike flood watermark photos, such taken photos should be handy for indicating historic peaks of the flood event. When overlapping photos were taken, applying Structure from Motion (SfM) data could be promising to extract peak elevations of such flood watermark photos.

For the first time, two mobile units were implemented for side-view SfM surveys. Applying these mobile units can speed up surveying processes in fields, especially for side-view surveys from roads. In this research, both mobile units were equipped with dual DSLR cameras to capture overlapping urban scenes. Two sets of Nikon® D5100 DSLR camera with 18-55 mm Nikkor™ lens were used (a zoomed-in box in Fig. 7-4a). These cameras maintained to be horizontal in alignment positions. The widest zoom function of lenses was set at 18 mm to retain a standard focal length. A shutter speed was set at 1/800 s and Vibration-Reduction (VR) technology of the lens was activated to prevent blurry photos captured by small involuntary movements of mobile units. Side-view photos were captured at 4,928 x 3,264 pixels (16 megapixels) to maximise resolutions of point cloud results. In particular, an advanced mobile unit (Fig. 7-4b) was equipped with a multi-frequency SPAN-SE™ GPS system with dual antenna and SPAN-CPT™ IMU system, whereas a simplified version of a mobile unit was operated without installing either GPS or IMU and occasionally used as a backup surveying unit. All mobile units maintain a groundspeed at 5.55 m s⁻¹ to prevent the shock capturing in motions and movements.

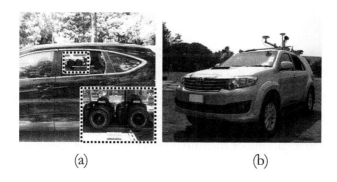

(a)   (b)

Fig. 7-4.   Two mobile units for side-view surveying: (a) a simplified version with the zoomed-in dual cameras on low right corner, (b) an advanced version with GPS and IMU systems

In the advanced mobile unit, another main operation was to measure elevations of street surfaces. When differential GPS (DGPS) approach was performed, recorded GPS signals can be post-processed for creating an absolute georeferenced location of the mobile unit. These GPS signals were synchronously received from both stationary and rover GPS units, in the meantime, IMU signals continuously were recorded orientations of the mobile unit. These orientations obtained from IMU system were then used to compensate some missing locations when GPS signals were weak. These absolute locations and elevations, aka POS data, were referred to the centroid of the IMU device (location 'A' in Fig. 7-5). Finally, street surface elevations (location 'A‡' in Fig. 7-5) can then be calculated from differences of such POS data relative to the vehicle height. These georeferenced locations and elevations of street surfaces can call as ground survey points (GSP).

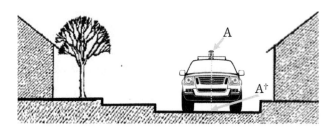

Fig. 7-5. A conceptual graphic of an advanced mobile unit using GPS and IMU for ground survey points (GSP) observation

For an application of measuring street elevations, an absolute accuracy of GPS with IMU system obtained from the mobile unit was ±0.02 m in horizontal and ±0.03 m in vertical (with post-processing mode), and referring to the georeferencing system of top-view LiDAR data (EPSG 32647: WGS84 UTM 47N). As a result, over 20 thousand locations of measured street elevations, called a ground survey points (GSPs), were then created.

7.3.2 Side-view SfM data processing

Even though applying fast shutter speed and Vibration-Reduction (VR) functions were cautiously set to stabilise cameras, some blurry photos were still captured due to shock capturing in platform motions and target-object movements. A subsequent removal of blurred photos was undertaken manually. Remaining sharp photos were selected and then ready for reconstructing point cloud using Structure from Motion (SfM) techniques. When these selected photos were converted and stored as a JPEG file format, making use of SfM technique can then be applied to indicate, match, and reconstruct 3D positions of features from consecutive overlapping 2D photos. SfM techniques use a feature detection method to indicate distinct reference points in each photo, using, e.g. the SiftGPU (Wu, 2007) open-source software. SiftGPU was further developed from the original software package SIFT (Lowe, 2004), implemented on an advanced multi-core Graphics Processing Unit (GPU). When these shared corresponding points of these features between overlapping photos were matched, they were then used to calculate extrinsic parameters. These parameters can be used to attempt to recover camera positions from correlative rotations, projections, and transformations of corresponding photo scenes by using 3D geometrical calculations. Bundle Adjustment (BA) method was used to improve accuracies of camera trajectories in calculations. BA can also minimise projection errors of camera tracking. Refined camera positions were next used to calculate sparse 3D point cloud data of target objects or each structure.

In this respect, multi-core bundle adjustment (Wu et al., 2011) in VisualSFM (Wu, 2011) open-source software were used rather than the ordinary Bundler (Snavely et al., 2006) software. Moreover, these sparse point cloud data were finally enhanced for reproducing denser SfM point cloud data by using an open-source software of Centre for Machine Perception - Multi-view Reconstruction (CMP-MVS) developed by Jancosek & Pajdla (2011). Even though much freely-alternative software is available (see Westoby et al., 2012), in this study, a commercial software of Agisoft™ PhotoScan™ was chosen due to it is easy to use and it has intuitive graphical user interface.

High-performance cloud computing system of SURFsara service (available for Dutch academic community) was applicable for side-view SfM point cloud processing for this study. Thirty nodes of the cloud system equip with Intel® Xeon® E5-2698 v3 of 32 processors at 2.7 GHz, and 256 GB of RAM. A virtual machine running on this system with 40 HPC cores and 64 GB of RAM was used, resulting over 600 million points of SfM point cloud were reconstructed from over 40 thousand overlapping 2D photo shots. A reconstructed side-view SfM point cloud was then georeferenced into the same coordinate system of top-view LiDAR data (EPSG 32647: WGS84 UTM 47N). The GSPs obtained from the mobile units were used as relatively georeferencing positions to side-view SfM point cloud in PhotoScan™. The details of the raw SfM data processing and registration were already given in Section 4.4.

The georeferenced SfM point cloud data were further extracted into two distinct key components, i.e. façade and low-level structure. While the façades represent the building fronts, walls, and opening pathways, low-level structures can represent the roads, and kerbs (Fig. 7-6a) features, following the process steps are given in Section 4.5. The selected façade point cloud can then be used to extract some hidden opening pathways. These openings were often, concealed in building fronts and walls, hidden under trees, or placed underneath overarching structures (Fig. 7-6b). These extracted urban features (i.e. low-level structures, walls, and opening pathways) of side-view SfM point cloud data were also rasterized at 5 m, 10 m, and 20 m grid resolution for creating side-view SfM-DEMs (i.e. Lowlevel-DEM and Opening-DEM) using ArcGIS®. Details of side-view SfM data extraction and rasterization process were given in Section 4.4.

(a)   (b)

Fig. 7-6.   An example of an opening concealed in arched entrance: (a) a photo taken from side view; (b) a side-view SfM point cloud, representing an opening wall and low-level structures (i.e. kerbs and roads)

7.4 Flood watermark extraction

Accuracies of historical flood watermark peaks observed in fields could improve post-flood analyses by making better understanding of flood magnitude and enabling knowledge insight into flood dynamics and processes happen in the past. This flood-related knowledge can also improve flood predictions, strengthen flood-protection measures, and make better resilience plans, especially for flood-prone cities. Even though flood peaks could also be analysed from time-series data recorded by systemic gauges, it may be more difficult to find such time-series records on land.

Some remote-sensed satellite images (e.g. RADARSAT, QuickBird, and IKONOS) can provide essential information to produce a flood extent map. Analysing such flood extent maps with high-resolution topographic data (e.g. LiDAR, SAR, and SRTM) could be adequate to estimate the peak of flood levels (Mason et al., 2012; García-Pintado et al., 2013). Even though the satellite images are regularly scanned and produced for each satellite for specific repetitive times, estimating such flood peaks using satellite images still has some limitations, due to these images may inconsistent with the exact time of the flood peak.

Unlike estimating the peak of flooding from the whole time-series records by gauging measurements, a post-flood analysis sounds to perform more practical and adequately used to estimate peaks of historical flooding. Estimating flood peaks from trash lines in watermarks remains the most common method that used in the post-flood analysis. Due to the localised nature of floodwater dynamics spatially results in its flood magnitudes not only for the long period flooding but also flash flooding (Bull et al., 2000; Hooke & Mant, 2000; Smith, 2014).

Even though peaks of historical flood watermarks were difficult to be detected from top views, such watermarks could be more easily observed from the ground (using conventional land surveys). However, such conventional surveys still use a lot of labour-intensive efforts. Several methods exist to obtain the peak of post-flood estimates; (Gaume & Borga, 2008) highlight the need to develop a robust, standardised method for calculating these estimates. In particular, an emerging Structure from Motion (SfM) technology should be suited for flood watermark extractions. Once the reference points have geographically been surveyed in advance, extracting flood watermarks from side-view SfM data may be promising.

In this study, the SfM technique was chosen as a side-view surveying for the flood watermark extraction, due to this technique is another outstanding remote sensing technology, which can reconstruct high-resolution topographic data from overlapping (digital) photos. A consumer grade digital camera or a camcorder as a side-view surveying tool can also be simply mounted on several surveying platforms, minimising accessibility gaps in technology resources.

7.4.1 Land surveying

Measuring peaks of flood watermarks can be made by using different techniques. The conventional levelling telescopes, total stations (Ballesteros Cánovas et al., 2011), GPS equipment (Sandercock & Hooke, 2010) or laser range finders (Denlinger et al., 2002) can be adequately used for this purpose. However, identification the peak of flood watermarks is often subject to considerable error (Marchi et al., 2009) and requires skill to identify the highest water level accurately. Such conventional surveys still use a lot of labour-intensive efforts.

In this case study, a conventional levelling telescope camera and a staff gauge were performed to measure four peak elevations of historical flood watermarks (Fig. 7-7). The levelling telescope SOKKIA™ B20 can measure the level of ±0.015 m vertical accuracy. The stationary GPS unit was used to create a ground control point (GCP) for relatively georeferencing to the measured watermark. The multi-frequency Leica Viva GNSS™ GS10 GPS was used to acquire the absolute coordinate location and elevation for each GCP with ±0.003 m horizontal and ±0.005 m vertical accuracies (DGPS with post-processing mode).

(a)   (b)

Fig. 7-7. An example of land surveying using (a) a levelling telescope camera and (b) staff gauge to measure an elevation of a flood watermark on the wall (location 'A' in Fig. 7-7a

7.4.2 Extracting flood watermarks from side-view SfM data

Even though a conventional land surveying is commonly used to measure the peak of flood watermarks, identifying and extracting such peaks are often subject to considerable error (Marchi et al., 2009) and requires skill to identify the highest water level accurately. Analysing multiple flood watermarks for an entire flooded area could minimise such errors (Gaume & Borga, 2008). However, when employing conventional land surveys to obtain such watermarks, it still uses a lot of labour-intensive efforts.

To date, a capability of mobile units in side-view surveying can reduce the time for capturing the overlapping photo shots. These shots of watermarks can be either simply captured, or highlighted with a high-visibility marker into the scenes to be easily identified in the point cloud afterwards. An inherently visual method in the SfM technique gives an advantage in its georeferenced side-view SfM point cloud, which can be applied for the flood watermark extractions. In this study, eleven flood watermarks were identified and extracted instantly from the georeferenced side-view SfM point cloud, providing the exact location with precise elevations of historical flood peaks (Fig. 7-8).

Fig. 7-8. The flood watermarks extracted in a coverage area of side-view survey using two mobile units and land surveying

The original images can be viewed once the sparse point cloud data were created and georeferencing processes were completed. Therefore, the post-flood analysis can be extended by identifying further high watermarks in the images. Given the focus on acquiring multiple high watermarks to increase the reliability of a post-flood survey an ability to conduct a careful desk-based interrogation of an entire area for flood watermarks are of substantial benefit to the survey. In this example, both side view of photo scenes and SfM point cloud can represent the same peak of flood watermarks, e.g. on walls (location 'A' in Fig. 7-9 to Fig. 7-11), and on electric poles (locations 'A' and 'B' in Fig. 7-12 and Fig. 7-13). Due to inundation of 2011 flood event in Ayutthaya Island lasted for months, these watermarks still can be found later years or even now.

(a)   (b)

Fig. 7-9.   An example of flood watermarks at 'A' on the gate: (a) captured from the photo and
(b) detected from side-view SfM point cloud (location 'M5' in Fig. 7-8)

(a)   (b)

Fig. 7-10. An example of flood watermarks at 'A' on the wall (a) captured from the photo and
(b) detected from side-view SfM point cloud (location 'M1' in Fig. 7-8)

(a)   (b)

Fig. 7-11. An example of flood watermarks at 'A' on the gate and wall (a) captured from the photo
and (b) detected from side-view SfM point cloud (location 'M7' in Fig. 7-8)

(a)   (b)

Fig. 7-12. An example of flood watermarks at 'A' and 'B' on the electric poles (a) captured from the photo and (b) detected from side-view SfM point cloud (location 'M6' in Fig. 7-8)

(a)   (b)

Fig. 7-13. An example of flood watermarks at 'A' and 'B' on the electric poles (a) captured from the photo and (b) detected from side-view SfM point cloud (location 'M8' in Fig. 7-8)

7.4.3 Comparison of flood watermark observations

Eleven flood watermarks as benchmark were observed by using conventional land surveys and other eleven watermarks were observed by using the more advanced side-view surveys as alternative tools. When using georeferenced side-view SfM point cloud observed by side-view surveys, these could be used to identify and extract the peak elevation of the flood watermark. The coefficient of determination was calculated to measure the agreement between the measured and extracted flood watermarks. The resulting quality was evaluated (Table 7-3) by comparing the residual (Diff) and percentile (%Diff) differentials.

Table 7-3 Comparison of eleven measured watermarks versus extracted watermarks using
side-view SfM data (%Diff from Eq. 3-1)

Watermark	Locations		Peaks			
	UTM-E	UTM-N	Measurement	Diff	%	Referencing
	(m)	(m)	(m msl)	(m)	Diff	
M1	669747	1587986	5.76	0.04	0.69	GCP
M2	670078	1588662	5.89	0.02	0.34	GCP
M3	669846	1588302	5.87	0.05	0.85	GCP
M4	669141	1588455	5.84	-0.04	0.68	GCP
M5	669159	1588011	5.88	0.07	1.19	GSP
M6	670406	1587939	5.86	0.08	1.37	GSP
M7	669376	1587597	5.88	-0.03	0.51	GSP
M8	670457	1587361	5.86	-0.05	0.85	GSP
M9	670493	1587163	5.84	-0.06	1.03	GSP
M10	668791	1586955	5.85	0.08	1.37	GSP
M11	668310	1586918	6.03	0.13	2.16	GSP

When applying GCPs as reference points (i.e. watermarks of M1 to M4), the comparison between the measured watermarks versus extracted watermarks can achieve the absolute elevation of 0.05 m or less than 0.85% of the difference. Whereas applying GSPs as reference points (i.e. watermarks of M5 to M11), it can achieve the absolute elevation of ±0.08 m or less than 1.37% of difference except for the extracted watermark at M11 with the elevation difference of +0.13 m. To determine depths of floodwaters observed by conventional land surveying, on one hand, the floodwater depths can be measured using conventional levelling telescope camera and a staff gauge to calculate the different between surfaces of a watermark related to a reference point (i.e. GCP or GSP). On the other hand, the floodwater depths extracted from side-view SfM data can be directly differentiated from the registered point cloud of side-view SfM data.

7.5 Creating multi-source views digital elevation model (MSV-DEM)

A quintessence of multi-source views (MSV) approach is that it can be used to minimise missing gaps in the single-view topographic data, i.e. top-view LiDAR data. This approach can provide an opportunity to substitute such missing data with other data obtained from different sources, different times, different resolutions, and different viewpoints. In this respect, a novel MSV topographic data were mainly focused on merging top-view LiDAR data with side-view SfM data. Top-view LiDAR point cloud and side-view SfM point cloud were simplified, separately. Scale factors of creating grid-based resolutions of 2D schematics should adequately represent all essences of key features (i.e. low-level structures, walls, and openings) and therefore translate them in their coarse or fine resolution model schematic cells.

In this section, top-view LiDAR point cloud and side-view SfM point cloud were extracted and simplified, separately (Section 7.2.2 and 7.2.3, resp). Merging of these different-source views DEMs (i.e. Building-DEM, Terrain-DEM, Lowlevel-DEM, and Opening-DEM) were then carried out in order to create new multi-source views digital elevation model (MSV-DEM) using ArcGIS®. This (iv) MSV-DEM (Fig. 7-14) was created at 5 m, 10, and 20 grid resolution and then translated to the same resolution for 2D model schematics using Grd2Mike tool in Mike Zero™. Details of merging multi-source view data were given in Section 5.3.

In this MSV-DEM, the missing vertical urban structures of top-view LiDAR data were substituted by side-view SfM data. From top-view data (i.e. a satellite image; see Fig. 7-15a), four openings concealed in the walls cannot easily be detected.

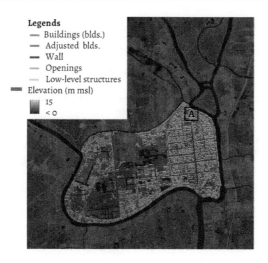

Fig. 7-14. An example of MSV-DEM at 5 m grid resolution with its legends on
top left corner ('A' zoomed-in area in Fig. 7-15)

Whereas both the novel MSV-DEM (Fig. 7-15b) and schematic map (Fig. 7-15c)
can clearly show some examples of these (missing) openings (locations 'A', 'B', 'C',
and 'D' in Fig. 7-15c).

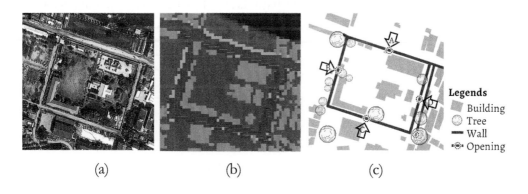

(a) (b) (c)

Fig. 7-15. (a) A high-resolution satellite image at Chantarakasem Palace National Museum,
Ayutthaya, Thailand (background: Google Earth™ 7.1.5.1557, 2014). (b) A zoomed-in example of
MSV-DEM of 5 m grid resolution. (c) A layout map with the legends on
the right (location 'A' in Fig. 7-14)

While closed walls in top-view LiDAR-DBM+ can clearly show in a 2D schematic map (Fig. 7-16a) using MIKE Animator Plus™. Whereas some openings concealed in the walls can explicitly show in plan view (locations 'A, B, C, and D' in Fig. 7-15c) and perspective view (locations 'A', 'B', 'C', and 'D' in Fig. 7-16b).

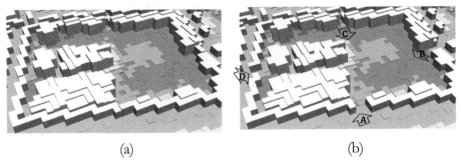

(a)   (b)

Fig. 7-16. Two example of two 2D schematic maps on (a) top-view LiDAR-DBM+ and (b) MSV-DEM with opening extractions, represented in the perspective view

7.6 Numerical modelling setups

A coupled 1D–2D model of Ayutthaya Island has been further enhanced to include the propagation of excess floodwaters from three main rivers (i.e. Chao Phraya River, Lopburi River, and Pasak River) and canals (Fig. 7-17). This was done in the MIKE FLOOD™ modelling system developed by DHI™. A 1D model of MIKE 11™ was used for simulating flows in rivers. The time-series containing discharges and water levels were used as external boundary conditions for the 1D model. These data were recorded by Royal Irrigation Department, Ministry of Agriculture and Cooperatives (RID/MOAC) during the 2011 flood event. The model instantiation and calibration were originally undertaken in the study of Keerakamolchai (2014) and the same model was further enhanced in the present study. The time series data simulated in the previous work, over the period of three months (September 1st to November 24th, 2011), was extracted and used as boundary conditions data for the purpose of the present work.

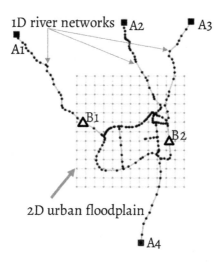

Fig. 7-17. A schematic of a coupled 1D-2D model
(1D river networks and 2D urban floodplain)

The time-series of the simulated discharges of: Chao Phraya River (Fig. 7-18a; location 'A1' in Fig. 7-17), Lopburi River (Fig. 7-18b; location 'A2' in Fig. 7-17), Pasak River (Fig. 7-18c; location 'A3' in Fig. 7-17), and water levels of Chao Phraya River (Fig. 7-18d; location 'A4' in Fig. 7-17) were used as boundary input data. A Manning's M friction coefficient of 40 was uniformly applied for all 1D river networks, followed the criteria defined by Chow (1959).

In 2D modelling setups using MIKE 21™, three raster-based topographic data of (i) a LiDAR-DTM, (ii) a LiDAR-DBM+, and (iii) a novel MSV-DEM in 5 m, 10 m, and 20 m grid resolutions were used as topographic input of a 2D urban floodplain. The Manning's M friction coefficient of 30, following Keerakamolchai (2014) study, was identically applied to the 2D urban floodplain for all these raster-based topographic data.

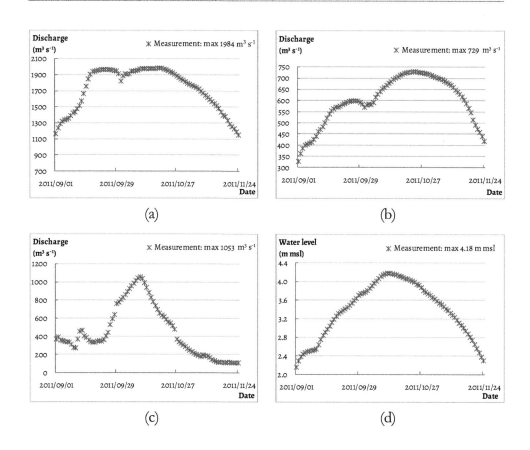

Fig. 7-18. Time series input at boundaries: (a) discharges of Chao Phraya River (location 'A1' in Fig. 7-17), (b) discharges of Lopburi River (location 'A2' in Fig. 7-17), (c) discharges of Pasak River (location 'A3' in Fig. 7-17), and (c) water levels of Chao Phraya River (location 'A4' in Fig. 7-17)

7.7 Results

7.7.1 Calibration of the models

The results of the coupled 1D-2D simulations were evaluated with other time-series of discharges and water levels at C.35 (location 'B1' in Fig. 7-17) and S.5 (location 'B2' in Fig. 7-17) water-gauge stations observed by RID. The coefficient of determination was calculated to measure the agreement between simulations and the daily discharges and water levels measured for about three months with considering to be avoided in too much force fitting manners. As a result, the quality of simulations was evaluated (Table 7-4) by summarising residual errors with the two water-gauge stations using two metrics, i.e. a coefficient of determination (R^2), and a root mean squared error (RMSE).

Table 7-4 Agreement between simulated results and measurements of discharges at C.35 station in Chao Phraya River and S.5 station in Pasak River (R^2 from Eq. 3-3and RMSE from Eq. 3-4)

Station	Discharge (m³ s-1)		Water level (m)	
	R2	RMSE	R2	RMSE
C35	0.99	23.83	0.97	0.16
S5	0.99	45.47	0.96	0.23

At the C.35 station in Chao Phraya River, the coefficient of determination (R^2) of discharges in the simulation was 0.99 with the root mean square error (RMSE) differed from the measurements by 23.83 m³ s⁻¹ (Fig. 7-19a).

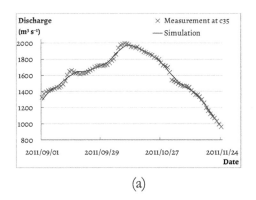
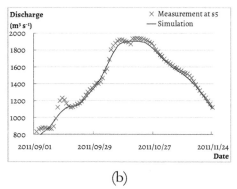

(a)   (b)

Fig. 7-19. A comparison between simulated results and measurements of discharges at (a) C.35 station in Chao Phraya River and (b) S.5 station in Pasak River

The R² of water level in the simulation was 0.97 with the RMSE differed from the measurements about 0.16 m msl (Fig. 7-20a). At the S.5 station on Pasak River, the R² of discharges in the simulation was 0.99 with the RMSE differed from the measurements by 45.47 m³ s⁻¹ (Fig. 7-19b). Whereas the R² of water level in the simulation was 0.96 with the RMSE differed from the measurements by 0.23 m msl (Fig. 7-20b).

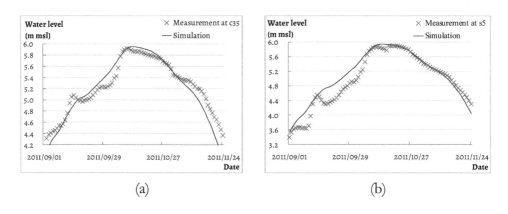

Fig. 7-20. A comparison between simulated results and measurements of water levels at (a) C.35 station in Chao Phraya River and (b) S.5 station in Pasak River

Some adjustment values in particular contexts of Manning's M friction coefficient ranges (i.e. 20 to 40) were tested and Manning's M 30 were chosen and uniformly apply for all 1D river networks. Some Manning's M values (i.e. 20, 30, and 40) were further tested for 2D floodplain area. Due to the fact that this study area is completely flat and flow evolutions of flooding are relatively slow, incorporating such Manning's M friction values ranging from 20 to 40 on LiDAR-DTM at 20 m resolution appear to show little effects. All simulated results of testing the different Manning's M friction values were also summarised. Therefore, in this case study, the Manning's M 40 were chosen and uniformly applied to all 2D floodplain area. However, adopting such values of Manning's roughness in other areas may vary and it should be handled with care. The more details in subject to a number of limitations when using of Manning's roughness were given in Smith (2014).

7.7.2 Comparison of 2D simulated floodwater levels

Eleven flood watermarks were observed by using conventional land surveying or extracted from side-view SfM data. In this case study, the eleven observations of flood watermarks as benchmark were used to evaluate the simulated results of the floodwater levels. The coefficient of determination was calculated to measure the agreement between the measured flood watermarks and simulations using different raster-based topographic data, i.e. (i) LiDAR-DTM, (ii) LiDAR-DBM+, and (iii) MSV-DEM in 5 m, 10 m, and 20 m grid resolutions. The resulting quality was evaluated (Table 7-5) by comparing the percentile (%Diff) differential.

When the simulated floodwater depths were relatively referred to the surface elevations of topographic input data (i.e. LiDAR-DTM, LiDAR-DBM+, and MSV-DEM), the floodwater levels of the simulated results can be determined. To compare the peaks of the simulated floodwater levels with benchmarked watermarks, both the peaks need to be defined in the same georeferencing system (EPSG 32647: WGS84 UTM 47N). Hence, the nearest grid of the simulated results, which related to the exact georeferencing location of the watermark was sampled and then evaluated. In particular, when applying the 20 m grid of LiDAR-DTM, LiDAR-DBM+, and MSV-DEM as input, the maximum differences between the measured flood watermarks and simulations were different of 8.45% at M1, 5.78% at M7, and 5.29% at M8. Whereas applying the 5 m grid of LiDAR-DTM, LiDAR-DBM+, and MSV-DEM as input shows the minimum differences at 1.02% at M6, 0.66% at M11, and 0.34% at M1. The most underestimated result was -0.49 m of difference at M1 when using the 20m LiDAR-DTM as input, and the most overestimated result was +0.15 m of difference at M11 when using the 20m LiDAR-DBM+ as input.

Table 7-5 Comparison of eleven measured watermarks (M1 to M11) versus simulated flood levels using different raster-based topographic data, i.e. (i) LiDAR-DTM, (ii) LiDAR-DBM+, and (iii) MSV-DEM in 5 m, 10 m, and 20 m grid resolutions (%Diff from Eq. 3-1)

	Locations M1		M2		M3	
	Level (m msl)	%Diff	Level (m msl)	%Diff	Level (m msl)	%Diff
Measured watermark	5.80		5.89		5.87	
1D-2D model using:						
20 m of LiDAR-DTM	5.31	8.45	5.41	8.15	5.41	7.84
20 m of LiDAR-DBM+	5.52	4.83	5.66	3.90	5.63	4.09
20 m of MSV-DEM	5.52	4.83	5.70	3.23	5.60	4.60
10 m of LiDAR-DTM	5.42	6.55	5.60	4.92	5.61	4.43
10 m of LiDAR-DBM+	5.60	3.45	5.70	3.23	5.70	2.90
10 m of MSV-DEM	5.58	3.79	5.78	1.87	5.68	3.24
5 m of LiDAR-DTM	5.51	5.00	5.77	2.04	5.67	3.41
5 m of LiDAR-DBM+	5.68	2.07	5.93	0.68	5.77	1.70
5 m of MSV-DEM	5.78	0.34	5.92	0.51	5.80	1.19
	M4		M5		M6	
	Level (m msl)	%Diff	Level (m msl)	%Diff	Level (m msl)	%Diff
Measured watermark	5.84		5.88		5.86	
1D-2D model using:						
20 m of LiDAR-DTM	5.35	8.39	5.45	7.31	5.50	6.14
20 m of LiDAR-DBM+	5.60	4.11	5.66	3.74	5.58	4.78
20 m of MSV-DEM	5.61	3.94	5.70	3.06	5.58	4.78
10 m of LiDAR-DTM	5.37	8.05	5.59	4.93	5.71	2.56
10 m of LiDAR-DBM+	5.65	3.25	5.74	2.38	5.77	1.54
10 m of MSV-DEM	5.70	2.40	5.71	2.89	5.72	2.39
5 m of LiDAR-DTM	5.51	5.65	5.61	4.59	5.80	1.02
5 m of LiDAR-DBM+	5.88	0.68	5.81	1.19	5.91	0.85
5 m of MSV-DEM	5.87	0.51	5.77	1.87	5.90	0.68

	Locations:					
	M7		M8		M9	
	Level (m msl)	%Diff	Level (m msl)	%Diff	Level (m msl)	%Diff
Measured watermark	5.88		5.86		5.84	
1D-2D model using:						
20 m of LiDAR-DTM	5.48	6.80	5.39	8.02	5.38	7.88
20 m of LiDAR-DBM+	5.54	5.78	5.57	4.95	5.57	4.62
20 m of MSV-DEM	5.60	4.76	5.55	5.29	5.56	4.79
10 m of LiDAR-DTM	5.59	4.93	5.50	6.14	5.44	6.85
10 m of LiDAR-DBM+	5.61	4.59	5.60	4.44	5.61	3.94
10 m of MSV-DEM	5.58	5.10	5.64	3.75	5.65	3.25
5 m of LiDAR-DTM	5.67	3.57	5.59	4.61	5.61	3.94
5 m of LiDAR-DBM+	5.74	2.38	5.77	1.54	5.77	1.20
5 m of MSV-DEM	5.78	1.70	5.79	1.19	5.80	0.68
	M10		M11			
	Level (m msl)	%Diff	Level (m msl)	%Diff		
Measured watermark	5.85		6.03			
1D-2D model using:						
20 m of LiDAR-DTM	5.49	6.15	5.85	2.99		
20 m of LiDAR-DBM+	5.55	5.13	6.18	2.49		
20 m of MSV-DEM	5.61	4.10	6.16	2.16		
10 m of LiDAR-DTM	5.66	3.25	5.90	2.16		
10 m of LiDAR-DBM+	5.69	2.74	6.11	1.33		
10 m of MSV-DEM	5.70	2.56	6.10	1.16		
5 m of LiDAR-DTM	5.79	1.03	6.14	1.82		
5 m of LiDAR-DBM+	5.81	0.68	6.07	0.66		
5 m of MSV-DEM	5.88	0.51	6.06	0.50		

7.7.3 Comparison of 2D simulated inundations

Comparison of inundations was undertaken for model results based on four different DEMs: (i) LiDAR-DSM, (ii) LiDAR-DTM, (iii) LiDAR-DBM+, and (iv) MSV-DEM.

The nature of flood processes in the case study area comprises both fluvial and pluvial floods which mainly result from higher flows coming from the three rivers (i.e. Chao Phraya, Lopburi, and Pasak Rives) as a result of the storms that occurred in the upstream areas of the Chao Phraya River Basin. Therefore, the upstream boundary conditions in the case study area include flows simulated in the larger river basin model (1D model) and they are fed through the coupled 1D-2D model of the Ayutthaya Island. The downstream boundary condition consists of the water level series at the Chao Phraya River. The time steps used for 1D-2D model simulations were set at 1 minute intervals. The maps produced depict maximum water levels computed across the 2D model domain.

In the case study work, the flood processes are driven by the slow moving river flows, and the terrain is almost flat throughout the region. Therefore, the velocities were found to be relatively low (around 0.1 to 0.3 m s^{-1}) and insignificant when compared to the magnitude of flood depths. Also, the duration of the flood event was for two months and our model simulations covered the period of three months. The event simulated is the actual flood event that took place from September until November 2011. In that event, and in the model results, the flood water remained for a period of two months. Hence, the following discussion addresses maximum computed flood depths. The inundation propagations and patterns can be estimated and compared when using four different types of DEMs: (i) LiDAR-DSM, (ii) LiDAR-DTM, (examples in Fig. 7-21a and b, resp) (iii) LiDAR-DBM+, and (iv) MSV-DEM, as input for urban flood simulation (examples in Fig. 7-22a and b, resp).

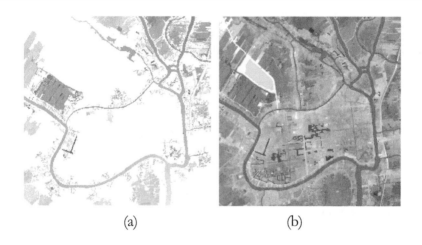

Fig. 7-21. Examples of simulated maximum flood depths using (a) LiDAR-DSM,
(b) LiDAR-DTM as input, represented at 20 m grid resolution

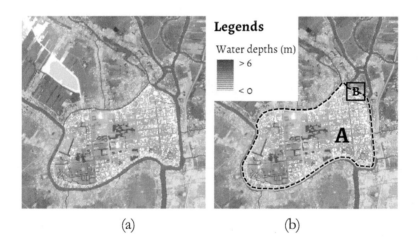

Fig. 7-22. Examples of simulated maximum flood depths using (a) LiDAR-DBM+,
(b) MSV-DEM as input, represented at 20 m grid resolution

These three different raster-based topographic data were also created at other different grid-based resolutions at 10 m and 5 m. By using these topographic data as input, the simulated results of the maximum flood depth and inundated areas of Ayutthaya Island (location 'A' in Fig. 7-22b) were then calculated to determine the flood volumes (Table 7-6).

Table 7-6 The simulated results of mean floodwater depths, inundation extents, and flood volumes using different raster-based topographic data, i.e. (ii) LiDAR-DTM, (iii) LiDAR-DBM+, and (iv) MSV-DEM in 5 m, 10 m, and 20 m grid resolutions

	Mean floodwater depth (m)	Inundation extent (m²)	Flood volume (m³)
1D-2D model using:			
20 m of LiDAR DTM			
20 m of LiDAR-DTM	1.96	7,602,403	14,900,710
20 m of LiDAR-DBM+	2.05	7,132,256	14,621,125
20 m of MSV-DEM	2.05	7,142,008	14,641,116
10 m of LiDAR-DTM	1.97	7,570,436	14,913,759
10 m of LiDAR-DBM+	2.09	7,079,050	14,795,215
10 m of MSV-DEM	2.08	7,128,905	14,828,122
5 m of LiDAR-DTM	1.99	7,469,180	14,863,668
5 m of LiDAR-DBM+	2.15	6,900,087	14,835,187
5 m of MSV-DEM	2.15	6,921,033	14,880,221

Moreover, the two raster-based topographic data concerning urban features, i.e. a LiDAR-DBM+ and a MSV-DEM, in 5 m grid resolution ('B' sub-region location in Fig. 7-22c) were also modelled.

Unlike using LiDAR-DTM (not concerning urban features) as input, the flood propagations were confined and freely flowed through opening walls and/or around buildings. When using the LiDAR-DBM+ as input, the simulated result of the maximum flood depth may be misrepresented some flood inundation (Fig. 7-23b), whereas using MSV-DEM as input, the result clearly showed the (missing) inundation area (Fig. 7-23c).

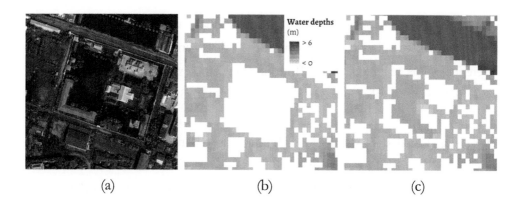

(a) (b) (c)

Fig. 7-23. A comparison of (a) a satellite flood-extent image (background: Google Earth™ 7.1.5.1557, 2014) and the simulated results of the maximum flood depth at 5 m grid resolutions, using (b) LiDAR-DBM+ and (c) a MSV-DEM as input

Even though simulation models were setup at fine temporal scale at 1 minute for each time step. However, simulated maximum flood depths seem to be practical to show final inundated areas, which can give reasonable distinguish examples when using different DEMs as input for 2D urban-flood models. Due to flood propagations in this case study had slowly developed during the entire flood event. Therefore, this (almost) stagnant situation may not reflect many dynamic characteristics of flood routeing processes.

7.8 Discussion

The side-view surveys had to be ended by the declaration of martial law on May 20, 2014. However, about 40 thousand overlapping shots and over 20 thousand GSPs had been surveyed using two mobile units with 64 man-hours in this study area. These side-view surveys were promisingly used to create high-resolution SfM point cloud, which was later used to extract eleven flood watermarks. During side-view surveys, none of observed (ground survey points) GSPs was not allowed for marking on public streets. Nearest locations (within radius distances 2.5 m of the street centre) were applied instead. However, these observed GSPs using the advance unit in side-view surveying were still promising to provide precise reference points of street centres. Both observed and extracted watermarks can be made, achieving the absolute elevation of less than 0.13 m vertical accuracies.

Processing DEM data with SfM technique allowed merging of side-view data, which in turn substituted some missing urban features into conventional top-view LiDAR data. In this way, a novel multi-source views digital elevation model, i.e. (iv) MSV-DEM, has been created. The flood model based on (iv) MSV-DEM produced the results which were more to reality (i.e. measurements). For example in 'B' sub-region, the floodwater flow directions are not only governed by the terrain characteristics, they are also governed by the vertical structures such as alleys and walls and possible openings that may be found inside such structures ('B' sub-region location in Fig. 7-22c; Fig. 7-23). With such data set, the floodwater can freely flow through the openings and create inundation, which is closer to reality.

Moreover, the key components, i.e. opening, pathway, and low-level structure cells, embedded into new (iv) MSV-DEM offer a new possibility for capturing important urban features into the domain of 2D hydrodynamic models. Furthermore, the parameter values of 2D models can be locally defined, for each type of particular key components, e.g. adjusting roughness coefficient values to the openings differed from the pathways, and adjusted accordingly during the calibration processes. Such roughness values can be locally defined and adjusted in calibration processes. The specific value can also be individually adjusted for each key component, e.g. adjusting roughness coefficient values to the openings differed from the pathways. However, roughness values at certain openings were not implemented in this case, due to the fact that the study area is entirely flatted and the flood evolutions in this area were relatively slow.

All flood simulated results presented here show that incorporation of urban features (buildings, walls, and openings) in a numerical model is an important aspect as such features can cause and can play a significant role in the shallow flow diversions, which commonly occurs in urban environments. When using (i) a LiDAR-DSM as topographic input data, most areas were dry and almost all floodwaters were mainly confined to the main channels. These high features (i.e. trees and overarching structures) were seen to behave as dykes. When using (ii) a LiDAR-DTM as input, the filtered data completely removed all high-urban features (buildings and high trees) with ground-flatted elevations. When applying (ii) LiDAR-DTMs as input, the simulated results only showed the floodwater flows spread over their flat terrain.

Due to all obstacles were filtered, the results seem to show a faster evolution of inundation, and it took 30 days for fully covered inundation. Results also show that applying LiDAR-DTMs slightly replicated the underestimate in floodwater levels than other two raster-based topographic data (i.e. LiDAR-DBM+ and MSV-DEM). Whereas analysing the results using (iii) LiDAR-DBM+ and (iv) MSV-DEM as input, it seems to show that their simulated results took the slower evolution of inundations.

Even though some complex urban features (e.g. alleys, openings, and low-level structures) can have a considerable effect on floodwater dynamics and predictions (Haile & Rientjes, 2005a; Boonya-aroonnet, 2008; Hunter et al., 2008b), all these urban features are usually not included in consideration for all conventional top-view LiDAR data, i.e. (i) LiDAR-DSM, (ii) LiDAR-DTM, and (iii) LiDAR-DBM+. Therefore, adopting such top-view topographic data as input for urban flood simulation still cannot replicate complex flow dynamics around such complex urban features.

7.9 Conclusions

The present paper describes the possibility of using SfM technique for extraction of flood watermarks. The survey campaign of the case study area, utilising two mobile units over the period of 64 man-hours, resulted in over forty thousand overlapping photo shots and twenty thousand GSPs. Adopting these mobile units for obtaining topographic data in side-view surveys is invaluable to reduce the time spent in this fieldwork. Such obtained side-view topographic data were used to create high-resolution SfM point cloud, which was later used to extract flood watermarks at eleven locations.

Extracted flood watermarks were then used as benchmark data set for verifying coupled 1D-2D urban flood models. Conventional top-view topographic data obtained from the aerial survey was used to create three top-view LiDAR topographic data: (i) LiDAR-DSM, (ii) LiDAR-DTM, and (iii) LiDAR-DBM+. The novel multi-source views (MSV) data processing method was developed for merging side-view SfM data with conventional top-view LiDAR data to supplement some missing urban features in conventional top-view data. The final result from that model is the novel multi-source views DEM (MSV-DEM). These four different types of DEMs were created at 5 m, 10 m, and 20 m grid resolutions. Overall, twelve DEMs were created and used to create twelve different 2D model domains which were then coupled with a 1D model and used for simulations of a 2011 flood event in Ayutthaya Island, Thailand.

Analysing all simulated results using different resolutions showed that higher resolution results can have more detail representations and better agreements with the measurements. This finding was logical since higher resolution topographic data can capture small discontinuities and depressions, which are otherwise smeared in lower (or coarser) simulation results. Due to the nearest grid of the simulated floodwater levels, which related to the exact georeferencing elevation and location of the benchmarked watermarks, can be better sampled in higher resolution raster-based topographic data.

Amongst four different DEMs, simulation results based on (iii) LiDAR-DBM+ and (iv) MSV-DEM data sets produced better quality urban flood maps. Even though the model results using (iii) LiDAR-DBM+ data were capable of representing more realistic flood flow routes, some results found to be slightly misrepresented. The presence of flow dynamics around complex urban features can be revealed and enhanced only when adopting MSV-DEM as input for urban flood simulation.

Overall, it can be concluded that new multi-source views (MSV) data obtained from side-view SfM data and conventional top-view LiDAR data, enables better representation of complex urban features for Ayutthaya's city. This new possibility for capturing important urban features can be embedded into the domain of 2D hydrodynamic models. Furthermore, the parameter values of 2D models can further be defined, for each type of particular key components locally, e.g. adjusting roughness coefficient values to the openings differed from the pathways, and adjusted accordingly during the calibration processes. Again, extracting floodwater marks from these new MSV data can also be invaluable for verifying 2D simulation results for the future.

CHAPTER 8
Recommendations for developing flood-protection measures: the case study of Ayutthaya, Thailand

The historic city of Ayutthaya, inscribed on the World Heritage list in 1991, is located in the central floodplain of Chao Phraya River Basin. Ayutthaya was subjected to the extreme flooding in autumn 2011, which can be counted as one of the worst flood disasters in Thai's history. Sixty-six of seventy-six provinces were inundated, of which Ayutthaya had the highest fatality of 97 deaths (BOE, 2011). In response to flooding in Ayutthaya Island, UNESCO Bangkok liaised with the Thai Ministry of Culture's Fine Arts Department (FAD) in mobilising national and international expertise for assessing flood damage extents and their causes, which occurred in the city centre of Ayutthaya. Results of two years' research proposed new flood-prevention approaches in cooperation with UNESCO-IHE, AIT, FAD, UNESCO Bangkok, HAII, and ADB. New approaches proposed to bypass floodwater from up streams along with enhancements of local flood-protection measures. Advances in urban flood models as hydroinformatics tools can play a significant role as scientific supports to new proposed flood-protection measures. In this chapter, identifying key problems of flooding in Ayutthaya case study is described in Section 8.1. Proposing regional and local flood-protection measures are expressed in Section 8.2. Establishments of flood-simulation results are shown in Section 8.3. Evaluations of new flood-protection measures are highlighted their benefits of using urban flood models as hydroinformatics supporting tools (Section 8.4). Some examples of public hearing and participations of local communities are shown (Section 8.5). Conclusions are given in Section 8.6.

8.1 Problem identification

Ayutthaya Island as a case study located in low and flat floodplain of Chao Phraya River Basin with average elevation of +3.50 m msl. The island is surrounded by three main rivers and one canal (Fig. 7-1). Ayutthaya City is susceptible to fluvial flooding (often occurred when river flows exceed its riverbanks), and it has experienced four devastating flood events in 1942, 1983, 1995, and 2011 (Table 7-2). Recently in the 2011 flood event, the entire Ayutthaya Island was inundated with water depths at some certain location exceeded two metres. This 2011 flood event in Ayutthaya was caused by excessive overflows from rivers (fluvial flooding) when a series of consequent tropical storms lasted for months in the northern parts of the country.

Based on several expert missions undertaken in November and December 2011, the International Council on Monuments and Sites (ICOMOS) and the International Centre for the Study of Preservation and Restoration of Cultural Property (ICCROM) recommended developing a flood disaster mitigation strategy for the Historic City of Ayutthaya. They indicated some main causes of severe flooding in Ayutthaya Island that the island located in low and flat terrains, extreme hydro-meteorological conditions of 2011, and lacking data supports. Amongst these supporting data, topographic data can be counted as utmost important information for supporting developments of risk mitigation policies and measures. Unfortunately, it seems to be hard to find decent details of topographic data during that time. However, for more than a year after the 2011 flood event, high-quality LiDAR topographic data were eventually ready, which distributed by Geo-informatics and Space Technology Development (GISTDA) surveyed by Japan International Cooperation Agency (JICA).

In this respect, advances in urban flood models were mainly used as hydro-informatics tools for local community participations in supporting and developing better flood-protection measures. However, assessments of community participation contexts for flood protection measures are beyond the scope of this research. Some examples of multidimensional approaches and their descriptions of economic, social, and cultural perspective analyses were already given and further discussed in a final report from ADB (2015).

8.2 Proposed flood-protection measures

After flooding in 2011, several flood mitigation plans were proposed for making better flood-protection measures. Amongst these plans, flood management plans for the whole Chao Phraya River Basin were purposed in cooperation with Office of National Economic and Social Development Board (NESDB), Royal Irrigation Department, Ministry of Agriculture and Cooperatives (RID/MOAC), Department of Water Resources, Ministry of Natural Resources and Environment (DWR/MNRE), and Japan International Cooperation Agency (JICA). Their study reviewed and evaluated several proposed measures for the Chao Phraya River Basin (see JICA et al., 2013). In order to alleviate flood risk on Ayutthaya Island, balancing of possible non-structure and structure measures were identified and their effects on the flood risk were considered (Table 8-1).

Table 8-1 Revision of proposed flood management plans (source by JICA et al., 2013)

Proposed measures	Measure revisions
Non-structure measures	
- Reforesting at upstream of river basin (C1)	Deforestation amplifies flood. Forest restoration requires continuous treatments over a prolonged period. The quantitative effects of flood mitigation produced by reforestation are not considered.
- Flood information management system	The flood management information system would play a critical role in the proper flood management. It is particularly emphasised that most of the damages in the factories can be minimised if the appropriate information on flooding and inundation is provided in a timely manner.

Structure measures

- Operation efficiency of existing dam (C7)	The operation of the existing dams during the 2011 flood was so effective to mitigate flood damages, since the Bhumibol and Sirikit Dams stored 12.1 billion m³ of floodwater. As the rule of dam operation was modified in February 2012, dam operation will have more flexibility to manage water resources with minimising flood damage as well as providing water for irrigation purpose. It is proposed that reservoir level should follow lower rule curve until the end of July, and from August, flood discharge should be stored in a reservoir with a maximum outflow of 210 m³ s-1 for Bhumibol Dam and 190 m³ s-1 for Sirikit Dam. If the proposed rule of dam operation is applied during the 2011 flood, the peak discharge at Nakhon Sawan could be reduced by 400 m³ s-1.
- Construction of new dams (C2)	Construction of new dams is highly encouraged, since it is effective for both flood mitigation and water utilisation for irrigation, especially in the tributary river basins. It is also promoted for response to climate change. However, dam sites currently identified cannot provide such large storage capacities as the Bhumibol and Sirikit Dams that the effectiveness of flood mitigation to the mainstream of Chao Phraya River is relatively limited.
- Improvement of retarding /retention areas (C4)	The areas with around 18,000 km², adjacent to river channels, currently have an important function to retard and retain floodwater. Therefore, it is crucial to preserve the areas not to lose the existing function by appropriately controlling land use. It is therefore recommended that land use regulations should be stipulated with considering scenarios such as excess flooding and climate change. To enhance the capacity of retarding floodwater, some measures such as the installation of gates and pumps can be taken. Those measures are useful to not only store floodwater but also to utilise the flood water for irrigation; however, the enhanced retarding effect is limited.
- East/West diversion channels (C6)	The diversion channels produce an enormous effect in reducing (i) water levels of the Chao Phraya River between Nakhon Sawan and Chai Nat, and (ii) inundation volumes flowing into adjacent retention/retarding areas. However, the effect of lowering water level produced by these diversion channels is fading away in the downstream stretch of Chao Phraya River close to the areas to be protected.

- Outer ring road diversion channel (C6)	The diversion channel has a particular effect in reducing water levels of (i) the Chao Phraya River from Ayutthaya to Bangkok, and (ii) the downstream of Pasak River. Hence, it is so effective to reduce the risk of dyke breaches along the areas to be protected.
- River channel improvement works (C5)	It is considered that the channel of the rivers lies between secondary dykes, not between water's edges along primary dykes, since the ordinary width of the stream cannot accommodate floodwater. It is crucial that lower and/or weaker stretches of secondary dykes should be identified and strengthened to prevent uncontrolled inundation. If dyke-raising work is conducted based on the primary dyke alignment, very high height of levee would be required, because river area enclosed with primary dyke is much smaller than secondary dyke.
- Ayutthaya Bypass Channel (C5)	The Ayutthaya Bypass Channel is one of the alternatives of river channel improvement works since it is extremely difficult to widen the river channel in the stretch between Bang Sai and Ayutthaya. The Bypass Channel has an effect in lowering the water levels of (i) the Chao Phraya River between Bang Sai and Ayutthaya, and (ii) the Pasak River. Hence, it is so effective to reduce the risk of dyke breaches along the areas to be protected.

Whenever information on flooding and proper flood modelling setups are provided promptly, simulated results could be applicable for developing and testing new proposed measures. In this respect, advances in flood models can play a significant role to help developing new proposed flood-protection measures. The recent research (see ADB, 2015) was adopted coupled 1D-2D urban flood models as hydroinformatics supporting tools in their study. New flood-prevention measures in responses to flooding in the Ayutthaya Island were purposed in cooperation with UNESCO-IHE, Asian Institute of Technology (AIT), Thai Ministry of Culture's Fine Arts Department (FAD) FAD, UNESCO Bangkok, Thai Ministry of Science and Technology's Hydro and Agro Informatics Institute (HAII), and Asian Development Bank (ADB).

8.2.1 Regional flood-protection measures

Even though the previous study by JICA et al. (2013) proposed countermeasures for major flood management plans for the whole Chao Phraya River Basin, possibilities for implementing whole plans are still unknown.

Amongst these countermeasures, the Ayutthaya Bypass Channel should be one of most feasible alternatives for improving channel capacities, which could improve an optimum capacity of a new Ayutthaya bypass channel up to 1,400 m³ s⁻¹, which was proposed following the ADB (2015) study. This regional flood-protection measure (location 'A' in Fig. 8-1) could result in lowering peaks of floodwater levels in Chao Phraya River. In this research, a maximum capacity of the new Ayutthaya Bypass Channel was set at 1,200 m³ s⁻¹ in order to prevent effects in dyke breaches, but still capable of reducing flood volumes in the Chao Phraya River between Bangsai and Ayutthaya Province.

Another potential alternative for river channel improvement works could be implemented on the eastern side of Chao Phraya River (e.g. Chainat-Pasak Canal, Raphiphat Canal, and Phra Ong Chao Chaiyanuchit Canal; location 'B' in Fig. 8-1), which related to the previous study by Panya Consultants (2012). Improvement works to these canals could be considered effective at reducing peaks of water levels and flood volumes with capacities of 1,000 m³ s⁻¹ for the Ayutthaya Island and nearby provinces located at the East of Chao Phraya River Basin.

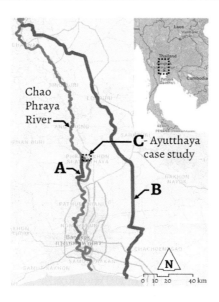

Fig. 8-1. Identification of regional flood-protection measures for
Ayutthaya case study (source by ADB, 2015)

8.2.2 Local flood-protection measures

From the ADB (2015) study, local protection measures were also applied to this study work. Some local flood-protection measures could potentially be implemented in Ayutthaya Island by (i) increasing dyke height, (ii) restoring ancient canals, and (iii) enhancing retention/detention ponds. Ayutthaya City, as an inland island, is surrounded by a ring road dyke (U-Thong Road), which runs along the edge of Ayutthaya Island. Existing approximate levels at +5.00 m msl of the ring road (Fig. 8-2) can serve as a dyke for the regional flood protection measure. From that existing elevation, this ring road dyke could be raised up to +6.70 m msl to create a dry polder for Ayutthaya City. The Province Office of Ayutthaya has sought an allocation of 550 million THB from the government to support their regional flood-protection measure, which involves building of over 12.5 km of dykes surrounding Ayutthaya Island.

Fig. 8-2. The existing elevation of U-Thong Road as a ring road dyke and
the road network in Ayutthaya Island (location 'C' in Fig. 8-1)

Another local flood-protection measure was to enhance the capacity of existing
retention ponds (wet ponds) and detention ponds (dry ponds) to retain the excess
floodwater in Ayutthaya's City.

In order to define appropriate locations for the ponds, three main criteria (i)
topography, (ii) land use, and (iii) existing drainage system were considerably
defined and analysed. Obviously, lowland areas of the island are the best location
for retention/detention ponds to collect such excess floodwaters through the
gravity. These lowland areas are vacant and feasible for the further flood-protection
development. The satellite images have also confirmed that these lowland areas
have a very low activity during the dry (red marked areas in Fig. 8-3a) and wet
season (red marked areas in Fig. 8-3b).

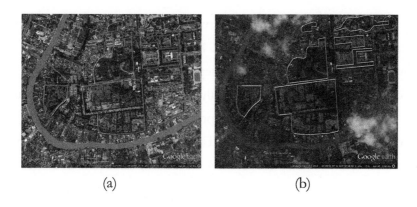

(a)   (b)

Fig. 8-3. The satellite images of lowland activities taken at the beginning of (a) the dry season on
December 5, 2013 (Google Earth™ 7.1.5.1557, 2013) and (b) wet season on
April 9, 2014 (Google Earth™ 7.1.5.1557, 2014)

Enhancing capacity of retention ponds could be made by excavating the more depths to existing retention ponds. Whereas increasing detention pond capacity could be created by dredging the berms and surrounding areas of such existing retention ponds. It was proposed to excavate the retention ponds from existing depths (Fig. 8-4a) down to 1.00 to 2.00 m or more depths (Fig. 8-4b). The locations and depths of retention and detention ponds were located nearby Bueng Phraram (location 'A' in Fig. 8-4b) and in Somdet Park (location 'B' in Fig. 8-4b).

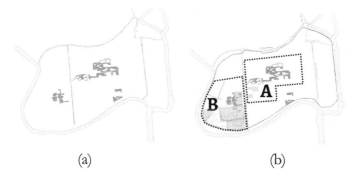

(a)   (b)

Fig. 8-4. Comparison of (a) existing ponds also, (b) proposed retention and detention ponds

The last local flood-protection measure was adopted from the concept idea of FAD development plan in the years 1994 and 2014, which proposed to restore Ayutthaya Water City as it was in ancient times. Reviving ancient canals was proposed as part of this strategy. Currently, most ancient canals are filled up with roads, with houses and buildings. The previous study by Jumsai Na Ayudhaya proposed a map of the ancient canals to The Siam Society in 1969 (Fig. 8-5a). The FAD used this reference map for reviving the ancient canal plan in 1997.

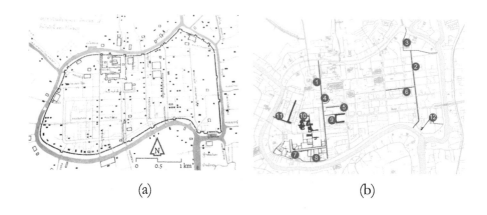

(a)   (b)

Fig. 8-5. The canals on Ayutthaya Island reviewed by (a) Jumsai's map and
(b) the dredge management plan of ancient canal and pond restoration
inside the island proposed by FAD

According to Jumsai's map (Fig. 8-5a), the total length of canals only in Ayutthaya Island was about 140 km, whereas the study by FAD shows that the existing canals only have 12.6 km left. It can be clearly demonstrated that existing canal networks are significantly reduced. Thereafter, FAD proposed the dredge management plan to restore twelve ancient canals in the island (Fig. 8-5b): (1) Thor Canal, (2) Makhamrieng Canal, (3) Maha Chai Canal, (4) Nakornban Canal, (5) Pa Tone Canal, (6) Bang Ian Canal, (7) the parallel canal to Lhek Road, (8) canals in Somdet Park, (9) canals around old municipal, (10) ponds in the culture village, (11) Rong Sura Canal, and (12) Wat Suwandaram Canal.

In the ADB's study, the existing situation, invasions, drainage capacity, and tourism
benefits were crucially concerned in order to evaluate the feasibility of ancient canal
restoration (Fig. 8-6). Even though ten ancient canals were evaluated, only seven
of them were potentially proposed for further development. Ignorant for the rest
of three canals, i.e. Pratu Khao Pluak Canal, Pratu Jin Canal, and Wat Suwandaram
Canal, because of the further implementing works no longer feasible for the current
situation.

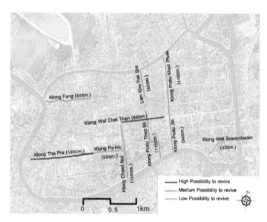

Fig. 8-6. Ancient canals potentially suitable for restoration (source by ADB, 2015)

In this study, the restorations of ancient canals and enhancement of
retention/detention ponds were proposed following the study by ADB (2015).
Capacity of the existing canals and ponds of 519,000 m³ could be increased more
to 980,000 m³, when the proposed ancient-canal restorations were built. Whereas
including with the enhancement of retention/detention ponds, the total capacity
should reach up to 1,498,000 m³.

8.3 Establishment of scenarios

In the field of hydroinformatics, a better quality of urban flood simulation may not
only be capable of replicating key behaviours of flooding in the past but also to

help modellers and decision makers better evaluating flood protection measures for the future. From the Chapter 6 and 7, amending geometries in MSV topographic data have already highlighted their capabilities to enhance the simulated results. In this respect, proposed flood-protection measures were simulated by modifying conventional 20 m resolution DTMs as input for the coupled 1D-2D modelling (Section 7.6). The scenarios of the case study were classified into four categories (Table 8-2): (i) the existing situation, (ii) the regional flood-protection measures, (iii) the local flood-protection measures, and (iv) the combination of regional and local flood-protection measures. All scenarios are based on the flood event of 2011, which can be counted as one of the most extreme flood event for the case study.

Table 8-2 Establishment of scenarios for flood protection measures

Scenarios	Dyke height (m msl)	Flood-protection measures
Existing situation		
Case 1	5.00	Without taking any newly proposed protection measure
Regional flood-protection measures		
Case 2-1	5.00	Ayutthaya Bypass Channel
Case 2-2	5.00	Ayutthaya Bypass Channel + East Chao Phraya Bypass Channel
Local flood-protection measures		
Case 3-1	6.30	Increasing dyke height
Case 3-2	6.40	Increasing dyke height
Case 3-3	6.60	Increasing dyke height
Case 3-4	6.70	Increasing dyke height
Case 3-5	6.30	Applying ponds and ancient canals + Increasing dyke height
Combined flood-protection measures		
Case 4-1	5.00	Ayutthaya Bypass Channel + Applying ponds and ancient canals
Case 4-2	5.80	Ayutthaya Bypass Channel + Increasing dyke height
Case 4-3	5.80	Ayutthaya Bypass Channel + Increasing dyke height + Applying ponds and ancient canals

8.4   Evaluation of the simulated measures

By applying urban flood models as hydroinformatics tools, proposed flood-protection measures can be evaluated. The floodwater depths and flood volumes (Table 8-3) can be calculated from such simulated results, which can also be used as input for flood-damage analyses afterwards.

Table 8-3 Simulation results of all scenarios under flooding event in 2011

Scenarios	Flood areas (km²)	Max depth (m)	Mean depth (m)	Flood volume (million m³)
Existing situation				
Case 1	7.60	5.20	1.96	14.90
Regional flood-protection measures				
Case 2-1	7.51	4.81	1.57	11.79
Case 2-2	0	0	0	0
Local flood-protection measures				
Case 3-1	5.74	4.24	0.87	4.99
Case 3-2	1.78	3.37	0.40	0.71
Case 3-3	0.83	2.43	0.25	0.21
Case 3-4	0.59	2.43	0.30	0.18
Case 3-5	5.43	5.19	0.90	4.89
Combined flood-protection measures				
Case 4-1	7.55	4.80	1.57	11.85
Case 4-2	7.57	5.00	1.11	8.40
Case 4-3	2.28	6.20	1.25	3.50

Moreover, all simulated results were given in four categories (i) the existing situation, (ii) regional protection measures, (iii) local protection measures, and (iv) combined regional and local flood-protection measures, as shown hereafter.

8.4.1 Existing situation

By adopting the coupled 1D-2D numerical modelling following the setups in Section 7.6, the existing scenario (Case 1) was modelled by using 20 m grid resolution of conventional top-view LiDAR DTM. This DTM can represent terrain elevations including the existing elevations of the U-Thong Road (ring road dyke) of Ayutthaya Island and its neighbouring territory. When adopted this DTM as input for the coupled 1D-2D flood modelling, the simulated results from the Case 1 show that the maximum flood depths were more than +6.00 m msl in rivers and ponds. Whereas most lowland areas inside the U-Thong ring road dyke (Fig. 8-2) was inundated with the approximate water depth of +2.00 m msl and flood volumes of nearly 15 million m^3.

8.4.2 Regional flood-protection measure

In Case 2-1 and Case 2-2 scenarios, such regional flood-protection measures involve the bypassing constructions for the Ayutthaya Bypass Channel and the East Chao Phraya Bypass Channel. The simulated results show that when only the Ayutthaya Bypass Channel (Case 2-1) was constructed, the flood volume can reduce from the existing situation (Case 1) by ~22%. Therefore, employing only this Ayutthaya Bypass Channel still cannot protect the Ayutthaya Island from flooding (e.g. the 2011 flood event). Whereas constructing both the Ayutthaya Bypass Channel and the East Chao Phraya Bypass Channel (Case 2-2) can eventually eliminate flooding and keep dry for all lowland areas inside the Ayutthaya Island.

8.4.3 Local flood-protection measures

When employing the U-Thong ring-road as a dyke, levelling the elevation heights of the ring road dyke could be increased up to +6.70 m msl (Case 3-4). As a result, when increasing the dyke height at +6.70 m msl, the simulated flooded areas (Case 3-4 in Fig. 8-7) can be significantly reduced to ~0.59 km² with the reduction of flood volumes down to ~0.18 million m³ (floodwater reduction of ~97%).

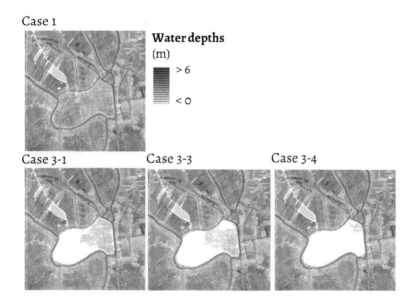

Fig. 8-7. Examples of maximum flood depths simulating the existing situation (Case 1) and the proposed dyke-height increasing measures (Case 3-1, 3-3, and 3-4, resp.)

8.4.4 Combined flood-protection measures

Implementing local protection measures show that they can significantly reduce a large number of floodwaters, when only the temporary or permanent dykes (Case 3-6) were built. It could be tough to get a better design in harmony with an urban context, which could be robust enough to handle the flood peaks like the 2011 flood event. While the constructions of bypass channels could protect the whole study area, such regional protection measures, which mentioned above are still in discussion and consideration process with an unpredictable decision from the government.

Thus, the combined protection measure between increasing dyke height and increasing ponds and canals capacity should be more effective for the case study. When incorporating bypass scenarios along with increasing dyke heights, we can minimise the heights of the ring road dyke down to +5.80 m msl (Case 4-5) in order to face with a moderate-inundated area of 7.05 km² with the flood volume of ~8.40 million m³ inside the dyke. By implementing one of the most possible regional-protection measures (Ayutthaya Bypass Channel) shows that the flood volume can be dramatically reduced down to 3.50 million m³ inside the dyke, which can be handled easily by installing movable pumps. In this respect, the local protection measures could also help to confine excess floodwaters within its canals and to convey them to the retention/detention ponds (Case 4-3 in Fig. 8-8).

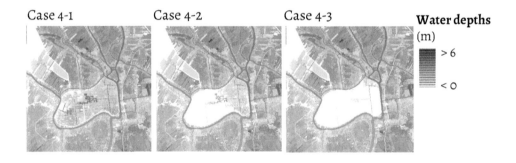

Fig. 8-8. Examples of maximum flood depths simulating the combined flood-protection measures (Case 4-1, 4-2, and 4-3)

8.5 Stakeholder preferences for flood-protection measures

By evaluating all measures from the simulated results, three local flood-protection measures were chosen: (1) raising of the U-Thong Road/Dyke by between 0.30 – 0.50 m from the existing elevation, and (2) creating a multi–functional landscape by upgrading the network of ponds and ancient canals, (3) enhancing the capacity of water gates and pump stations. These chosen local protection measures were further investigated for the preferences of stakeholders, which some examples were given elsewhere (e.g. Abbott & Vojinović, 2010a; Abbott & Vojinović, 2010b; Abbott & Vojinović, 2014; Vojinović et al., 2016).

8.5.1 Community preferences

During the community workshop on April 29th and May 22nd, 2014, in collaboration with Ayutthaya Municipality, the invitation letters of the workshop were sent to 33 communities in the island. By using topographic maps together with simulated flood maps, it was considered that quality of supporting hydroinformatics data using during the workshop could help the focus group to bring and reflect the preference of the community as a whole. The opinion of Community members towards the set of measures was positive. There were no significant objections to any measure. During the discussion, more concerns were raised due to developing the plan for flood protection on the island that could be summarised as follows;

1. *Construct new and improve the existing water gate*
- *Repair and improve the water gate at northern part of Mahachai Canal (location '1.1' in Fig. 8-9)*
- *Repair and improve the water gate at eastern part of Mahachai Canal (location '1.2' in Fig. 8-9)*
- *Build the water gate at Wat Sra Mon Tol Canal (location '1.3' in Fig. 8-9)*
- *Build the water gate at eastern part of Mueng Canal (location '1.4' in Fig. 8-9)*
- *Build the water gate at western part of Mueng Canal (location '1.5' in Fig. 8-9)*
2. *Construct a new permanent dyke at Hua Lor Market to prevent floodwaters overtopping the U-Thong ring road dyke (location '2' in Fig. 8-9)*
3. *Elevating up the dyke height from Prasat Temple to behind the Hua Lor Market (location '3' in Fig. 8-9)*
4. *Construct a new permanent dyke from Wat Sra Mon Tol Bridge to Soi U-Thong 27 (location '4' in Fig. 8-9)*

5. *Improve drainage system at Lhek Road to reduce floodwaters in Wangchai, Rimwang and Pratumnuk communities (location '5' in Fig. 8-9)*

6. *Improve the capacity of Mahachai Canal to retain the more excess of floodwaters (location '6' in Fig. 8-9)*

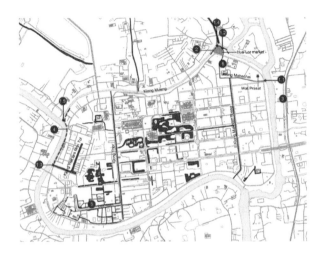

Fig. 8-9. Map of flood-protection measures proposed by communities and stakeholders (source by ADB, 2015)

8.5.2 Stakeholder preferences

In the consultation meeting and DRM workshop on May 27th to 31st, 2014, questionnaire sheets were distributed to all participants. Twenty organizations as the stakeholders were participated and carefully fulfilled the questionnaires, which elaborated by FAD, Ayutthaya Municipality, Phra Nakhon Si Ayutthaya Public Works And Town and Country Planning Office, Ban Chiang National Museum, Village Health Volunteer, Tourism Authority of Thailand, ICOMOS Thai Association and Faculty of Architecture and Planning, King Mongkut's University of Technology.

The stakeholders firstly preferred to enhance the capacity of the water gates and pumping stations. One of the major reasons that are enhancing the capacity of water gate and pump station was highly preferable, due to the existing conditions and infrastructure. Compared with other measures, it will be much quicker and easier to start implementing the plan. Two other pumping stations at Makamreang Canal and Thor Canal were also mentioned as a first priority. The second preference was related to the creation of a multifunctional landscape, by upgrading the network of ancient canals and ponds. Third preference was to increase the dyke heights up to 0.3 – 0.5 m from the existing elevations by using the U-Thong Road as a ring road dyke. The majority of participants from the FAD also gave high preference of creating a multifunctional landscape by upgrading the network of ponds and ancient canals since the work will directly complement the heritage works of FAD (Keerakamolchai, 2014).

8.6 Conclusions

Flood extents and their flood related issues of the case study – Ayutthaya Island were investigated by using coupled 1D-2D numerical flood models as hydroinformatics supporting tools. All proposed flood-protection measures in both the regional and local flood protection measures were simulated, evaluated, and finally represented for the public participations. The simulated flood maps with the different proposed measures were demonstrated that such hydroinformatics supporting tools and their valuable information could enhance in accord with the perceived flood risk to both communities and stakeholders. Thus, the combined flood-protection measures (i.e. the Ayutthaya Bypass Channel as the regional flood protection measure together with the increasing dyke height, reconstructing ancient canals, and enhancing pond capacities as the local flood-protection measures) was chosen as the most effective flood protection measure. Even though all previous flood protection measure was practically used quite some time in the past, the new proposed flood-protection measures of this study should bring more effectiveness of flood mitigations for Ayutthaya Island for the future.

In this section, floodwater depths and flood inundation patterns may have slowly changed during this very long time scale. Despite using a great detail with highest resolutions for model simulation, top-view LiDAR-DTMs were mainly used to achieve reasonable flood inundation maps for further analysing flood depth damages. However, in the initial stages of flood events, floodwaters could flow around each building until they reach openings. Then waters will enter buildings or flow through them. When susceptible openings should be taken into account in urban flood modelling (Chapter 7). The flood simulation results should reveal and indicate each of which building are prone to flooding and help for indicating locations of each susceptible entrance.

Overall aims of this chapter concentrate mainly on the local and regional flood-protection measures, this study did not focus on awareness of public perception, environment, and aesthetic issues. These issues are out of scope of this research and they are still controversial which could lead to more debate in social science communities elsewhere. However, it must be highlighted that the more detail aspects would require further studies for coupling such issue. Further studies should be carefully considered all these effects in details with a full-scale of engineering feasibilities to meet most requirements for stakeholders before any firm conclusion could be drawn.

CHAPTER 9
Outlook of multi-view surveys and applications

Benefits and applications of new multi-source views (MSV) topographic data mentioned in previous chapters were mainly obtained from side-view surveys using SfM techniques and top-view surveys using LiDAR systems. In this chapter, outlook of obtaining topographic data from different viewpoints are explored in Section 9.1. Three advanced small-format surveying platforms are further examined, i.e. unmanned aerial vehicles (UAV) for aerial-based surveys (Section 9.2); mobile mapping systems (MMS) for land-based surveys (Section 9.3); and unmanned surface vehicles (USV) for aquatic-based surveys (Section 9.4). Large improvements in topographic data acquisitions can also be found in many smaller and cheaper surveying tools. Some surveying tools these days can be multi-installed into the same small-format platforms. Amongst these tools, some new cameras can reveal their hidden capability for performing in low-light situations, which could be helpful for some difficult surveying conditions (Section 9.5). Furthermore, vast improvements in new numerical model engines can also provide a great opportunity for better analysing complex flood situations (Section 9.6), and more advances in high-performance computing are now capable of simulating such complex floods much faster than before (Section 9.7).

9.1 Obtaining topographic data from different views

Since remote sensing technologies have emerged, quality of topographic data has dramatically improved. Observing the Earth's scenes are typically obtained only from one source with one viewpoint. Earth's surface observations relative to the horizon can be categorised in three main different viewpoints: (i) a top view, (ii) an oblique view, and (iii) a side view (Fig. 9-1).

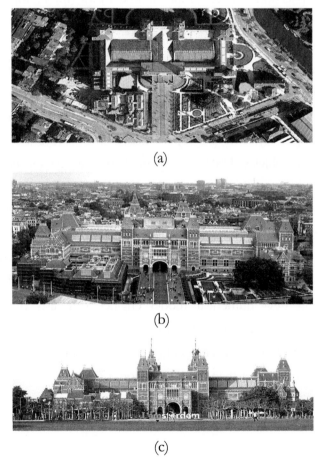

(a)

(b)

(c)

Fig. 9-1. Three different views of the iconic landmark – Rijksmuseum, the Netherlands: (a) a top-view image obtained from DigitalGlobe Satellite (Google Earth™ 7.1.5.1557, 2016b); (b) an oblique-view kite photo taken by Elbers (2013); (c) a side-view photo taken by Catarinella (

Amongst of these three main different viewpoints, (i) top view can be counted as the most popular use in the Earth's observations. Due to obtaining top-view topographic data from aerial-based surveys are much faster than obtaining data from land-based surveys especially for remote areas. Many small-format aircrafts these days are capable of programmed flight-path and self-navigation systems, which could be operated as advanced small-format unmanned aerial vehicles (UAV) for alternative aerial-based surveys.

Other (ii) two oblique and low-oblique views can be obtained from aircrafts and also from a rooftop of skyscrapers (Fig. 9-1b). Distinction between of both oblique views can be categorised by appearance of the horizon. In the oblique views, the horizon typically appears in the upper part the scene. Whereas in low-oblique views, they have no horizon captured in the scene. Many oblique scenes of urban landscapes, skyline, and their complex surroundings of urban areas are attractive and beautiful. Even though these fantastic scenes can represent the landscapes from above, substantial distortions are still embedded in these oblique scenes (Ham & Curtis, 1960). When such oblique scenes are used for topographic data processing, such oblique-view data (Fig. 9-2) are commonly contained with a high degree of geometric errors and can be difficult for geographical and geomatical corrections. Therefore, these oblique-view surveys have less attention for obtaining high quality topographic data.

For (iii) side view, it could be perceived as almost similar as high-oblique viewpoints, due to the horizon commonly appears in the middle or in lower part of the scene (Fig. 9-1c). This horizon should be found at a level position rather than tilted at an odd angle. These horizon appearances are quite common for the people to seeing the world from the ground (Wildi, 2006).

Fig. 9-2. An example of oblique-view SfM point cloud obtained from
rooftop of the Nieuwe Kerk (New Church), Delft, the Netherlands

In land-based surveys, despite performing stationary surveys, installing surveying
tools in mobile vehicles allows surveyors to perform their surveying missions much
faster. Mobile mapping systems (MMS) could be adequately used for side-view
topographic data acquisition.

Apart from land-based surveys, the concept of mobile units is not only limited to
land-based surveys, but it could also be applied to aquatic-based surveys.
Nowadays, some aquatic-based platforms can be operated remotely. These aquatic-
based platforms can be applied to both side-view surveys (Fig. 9-3) and underwater
surveys (e.g. using echo sounder for bathymetry data acquisition), see some
examples in Vojinović et al. (2013). When echo sounder systems are installed
onboard, such aquatic-based units will be ready for observing bathymetries of
rivers, canals, and ponds (bathymetries). Decent details of such bathymetry data are
necessary for estimating topographic profiles and capacities of water bodies which
crucial for flood schematizations and their simulation setups.

Fig. 9-3. An example of a side-view scene captured from a boat in
the Pasak River, Ayutthaya, Thailand

Recent advances in remote sensing technologies for Earth's observations these days can be used for achieving topographic data from different sources, different viewpoints, different times, and from the combination thereof. Also, some examples are further examined hereafter.

9.2 Unmanned aerial vehicle (UAV)

Several airborne platforms (e.g. aeroplanes, helicopters, hot-air balloons, large blimps, sailplanes, and including spaceborne platforms) are commonly used in aerial-based surveys for a long time already. However, these conventional airborne platforms use a lot of investment costs and necessarily need well-trained or professional pilots for each surveying operation. Despite employing such conventional aircrafts, many small-format aircrafts have recently been used and operated as alternative tools for aerial-based surveys. Investment costs for these small-format aircrafts are much cheaper than the big and conventional aircrafts. In particular, these small aircrafts can range from only a few USD 100 for a small simple aircraft, and up to USD 10,000 for a larger and more sophisticated aircraft (Aber et al., 2010). Many small-format aircrafts are high portability, rapid setup, and easy to use in the field. Owing to this, adopting such small-format aircrafts can be logistically possible for many applications (Aber et al., 2010). When capabilities of programmed flight paths and self-navigations are enabled in such small-format aircrafts, they can be used as small-format unmanned aerial vehicles (UAV). Benefits of adopting these small-format UAVs can be highlighted in the situation when employing conventional aerial-based surveys would be too risky, impractical, or even impossible to be operated safely. Therefore, employing small-format UAV can be counted as next generation of aerial-based surveys for several applications.

Most small-format UAV could compromise with lightweight loadings. Due to these UAVs are commonly dedicated only for aerial-based surveying purposes and they often contain surveying tools without concerning any pilot weight. Thankful for huge improvements in surveying technologies, surveying tools these days become smaller and cheaper. These small and compact surveying tools can be multi-installed into the same small-format UAV.

Amongst of these surveying tools, cameras have long been used in airborne platforms nearly since the first flying object was invented. New low-cost and lightweight digital cameras together with aircrafts could be a good combination for most aerial-based surveyors. Many consumer-grade digital camera costs can range from a few USD 100 to several USD 1,000 and they are much cheaper than large-format aerial cameras, which cost starting from several USD 100,000 (Malin & Light, 2007).

Apart from this research, a new small-format UAV was developed in Hydro and Agro Informatics Institute, Ministry of Science and Technologies (HAII/MOST), Thailand for other aerial-based surveying missions. In this small-format UAV, a small digital camera was mounted underneath the platform (Fig. 9-4), without any propeller obstacle and no need for synchronising motor modifies. Two nickel-metal-hydride (NiMH) batteries were used as a power source, which can be recharged quickly. This UAV used fix-wings, which enable to have stable flights. A simplified version of controlling flight directions used rudders and elevators without helping of ailerons. Two radio channels were opened, one for an aircraft controller and another for operating a camera shutter using micro-servo. This UAV was designed to allow some aircraft parts to be re-packable and portable in a medium-size storage box, such as a golf travel case.

Fig. 9-4. An example of a fix wing UAV with a camera installation

Almost three weeks were spent on aircraft trainings. All aircraft orientations, acceleration speeds, flight altitudes, over controls, and pre-flight checks also wind and weather conditions were carefully concerned for successful model aircraft piloting following Graves (2007) notes. However, becoming a master in take-offs, landings, and flights were not easy and required a lot of experiences for controlling aircraft properly. Enabling programmed flight path and self-navigation capabilities could help and then they were further considered for upgrading to a new small-format UAV with programmed flight-path and navigating systems.

When global positioning system (GPS) and inertial navigation system (INS) units were mounted onboard, another radio channel was provided for transferring all information of flight paths, orientations, and directions from the aircraft back to the ground controller. These continuous two-way radio communications synchronously updated the flight paths and current positions of the aircraft to be visualised in Google Earth base-map interface at almost real-time. With these two-way communications, deviating or even enforcing the aircraft were possible. Ground speeds of the aircraft were set at 45 to 70 km h^{-1} and lower ground speeds were applied when wind speed increased. Flight paths must be arranged in parallel to the wind direction. Therefore, the ground speeds can be varied between ongoing and returning paths. The remote-controllable functions were still maintained in this new small-format UAV in order to be toggled between the programmed autopilot and the handheld radio controller. Even though this new UAV are fully functional of autonomous controls, operating this UAV was recommended to approach take-off and landing phase manually for safety reasons.

When this new small-format UAV was ready, a digital camera Canon IXUS 230 HS was then mounted onboard as a surveying tool. Some camera setups were also tested at lower flying heights for maintaining 60% overlapping areas between consecutive photo shots. The shutter speed at 1/1,000 s, lens exposure at f3, and ISO at 320 were all applied as regular pre-set of the camera. An onboard computer in the aircraft used this pre-set to approach the camera shutter for capturing each photo shot (Fig. 9-5).

Fig. 9-5. An example of aerial photo scene captured from the UAV

All GPS and INS data were recorded into the onboard computer. When the camera shutter was activated, the ground geo-location the currently captured photo was logged into EXIF metadata of each JPEG file. These EXIF metadata are also helpful for categorising photo selections, and photogrammetric processing and analyses. Such geo-located JPEG files of all captured photos were then ready for visualising in geo-referenced map (Fig. 9-6).

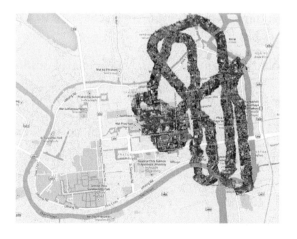

Fig. 9-6. Examples of georeferenced aerial-photo shots taken
along a programmed flight path

However, some aerial photos needed to be cross-checked whenever these photo shots did not match with the geo-referenced map. At this point, our current UAV missions were used only for capturing some snapshots or recordings some video scenes. Whenever this UAV setups are more stable and reliable, it will be used in full functional aerial-base surveys for top-view topographic data acquisitions for the future missions.

9.3 Mobile mapping system (MMS)

Nowadays, may surveying tools can be installed and operated on mobile mapping systems (MMS) or mobile units. Several advanced surveying tools can be mounted to almost any vehicle for many land-based surveying purposes.

In this research, advanced mobile mapping systems (MMS) were equipped with Global Navigation Satellite System (GNSS) and inertial measurement unit (IMU). Therefore, positioning of the vehicle can be correctly identified in a short time. This MMS was also integrated with a laser scanner system and two digital single-lenses reflex (DSLR) cameras as surveying tools.

Details of topographic data can be precisely observed. Locations of each urban feature and their key component details were recorded. These valuable recorded data were further analysed and processed later in the office. Our current MMS (Fig. 9-7) is still in the research phase, and it will be ready for real surveying operations in the near further missions.

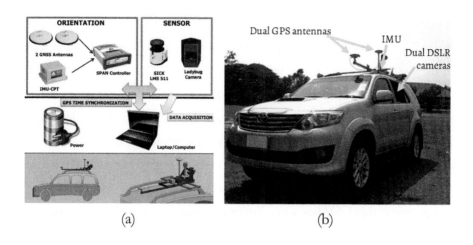

(a)   (b)

Fig. 9-7. An example of (a) surveying tools: dual DGPS, IMU, Dual DSLR cameras, and a laser scanner system (source by Boonya-aroonnet et al., 2015), which installed on (b) the MMS

9.4 Unmanned surface vehicle (USV)

Autonomous surface vehicles (ASV) or unmanned surface vehicles (USV) are commonly operated on open-surface waters, but they could be valuable for both side-view surveying and underwater surveying applications. Movability, portability, and capability of multi-installing surveying tools in USVs can bring new benefits to many surveying purposes, which could be far more flexible than stationary, moored, or drifting weather buoys. These unmanned vehicles can be operated on water surfaces without any crew onboard. Some vehicles allow the operator able to toggle between programmed route and hand-held controls. When employing USV for aquatic-based surveys, operating such aquatic-based units is of course far cheaper than employing conventional surveying ships and research vessels.

New USV tends to become cheaper and smaller. Owing to this, it should have more flexible than commercial-ship contributions. Key capabilities in such aquatic-based surveys can be emphasised in situations that neither aerial-based surveys nor land-based surveys would be impractical or impossible to complete surveying missions. Apart from this research, a new small-format USV was developed in HAII/MOST for other aquatic-based surveying projects. In this small-format USV, GPS system, IMU system, and echo-sounder device were installed on a kayak platform. Automated route controls were also applied for navigating this USV (Fig. 9-8). In practical, this USV has some limitations that they were not fully automated in strong current in canals. When these USV setups were ready, all information of routes and directions of each surveying path can synchronously be recorded along with bathymetries using echo sounder systems. Such echo system can provide the water depths between USV and underwater ground surfaces. The water surfaces were referred to the mean sea level (msl). Then these water surfaces calculated using DGPS and IMU data and water depths from echo-sounder data were used for determining elevations of bathymetries later in the office.

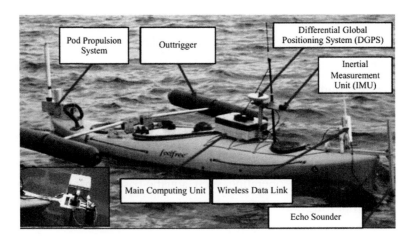

Fig. 9-8. An example of USV equipped with DGPS, IMU, and echo sounder system on a kayak platform (source by Boonya-aroonnet et al., 2015)

Many surveying missions in 2011 were applying this USV for observing canal bathymetries, e.g. Klong Lad Phrao, Klong San Saeb, and Klong Mahanak. Those observed canal bathymetries were important for calculating canal capacities. Such pond, floodplain, and canal capacities were then used for detaining, retaining, and conveying floodwaters before entering the capital city – Bangkok, and after floods for draining such excess floodwaters directly through drainage canals, tunnels, rivers, and finally to the sea. During surveying missions, we found that a number of shallows and deposited sediments were located at almost all junctions of canals also most places where the expressway foundations situated in the canals. Our surveyed results can swiftly point out and precisely identify these exact deposited-sediment locations to be later dredged for improving canal capacities.

9.5 Night vision cameras for enhancing side-view surveys

Whenever the normal daylight photography cannot be taken or even impossible to be performed (e.g. observing some buildings intercepted by crowd activities through the day), capturing photo shots using night-vision surveys could be another good alternative option.

Almost all new digital camera sensors are sensitive for not only visible spectrums but also near infrared (NIR) spectrums. Capabilities of taking photo shots in NIR spectrums can be currently found in some new digital cameras. Typically, a blocking filter (hot mirror) is placed in front of camera sensors, allowing only visible spectrums passing through, but prevent NIR spectrums. Due to the intervened NIR spectrums in a NIR photo shot are odd and different from a standard colour photo shot for common human perceptions.

Apart from this research, two Nikon D5100 digital single-lens reflex (DSLR) cameras were modified for night-vision surveys. By removing this hot mirror, the NIR spectral sensitivities of the sensor can be activated. Then, a blocking filter can be replaced by either an infrared (visible light blocking) filter for pure NIR photography or a clear filter for almost-full spectrum photography. In this respect, the clear filter was applied for our cameras due to, in each photo shot, it will capture almost all photon spectrums of both primary visible spectrums and NIR spectrums. However, when we applied these cameras at low light conditions, the almost night scenes were still hard to capture a good photo shot. When applying wide apertures with high ISO and moderate shutter speed, the captured photos were slightly grainy and dark. Even though applying lower shutter speed can make some better and brighter photos, this approach could be possible when the vehicle move very smooth and slow, or even parked. Four infrared spotlights (Fig. 9-9b) were installed on the same platform of these two NIR cameras (Fig. 9-9a). We found that these setups were promising for capturing photo scenes at very-low-light situations or even at night. Such infrared spotlights are less bright than normal spotlights. Activating such infrared spotlights for night-vision surveys may have little nuisance to night-life civilians.

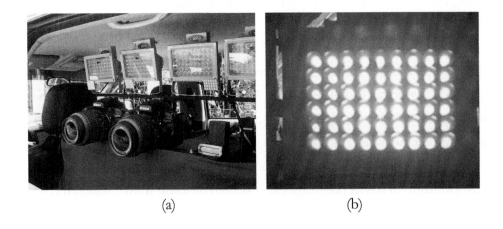

(a)   (b)

Fig. 9-9. An example of night-vision side-view surveys using
(a) dual NID cameras with (b) infrared spotlights

9.6 Enhancing 2D model schematics

Evaluating 2D urban flood model performances commonly relies on two major indicators: accuracy and efficiency. While (i) considering terms in the governing equations for describing detail behaviours of flood flow dynamics, (ii) applying higher order precisions of numerical methods for minimising inaccuracies and errors of discretization, or (iii) adopting finer spatial and temporal resolutions and enhancing 2D model schematics for replicating more key components of local urban features, could improve higher accuracy for 2D models. Whereas (iv) neglecting insignificant terms for reducing the equation complexities, (v) simplifying 2D model schemes for speeding numerical solving, (vi) reducing size or dimensionality of modelling domain for minimising the computational time, or (vii) employing faster hardware for shortening the computing time. These two major indicators, i.e. accuracy (i to iii) and efficiency (iv to vii) are commonly in conflict with one to another (Chen et al., 2012a). Almost all chapters mentioned earlier are mainly focused on improving accuracy of 2D urban flood model by multi-source view (MSV) method for enhancing 2D model schematics, other accuracy improvements should be found elsewhere. But some efficiency enhancements (i.e. v and vii) are further discussed hereafter ('vii' in Section 9.8).

Even though applying high-resolution topographic data as input for 2D models could represent a more detailed flow dynamics in complex urban areas, computational cost for simulating such fine resolution 2D models could exponentially rise as the resolution goes finer. For (v) simplifying 2D model scheme should reduce the computing efforts for 2D flood-model simulations. The traditional grid adjustments commonly used average elevations of a fine grid to create a new coarse-grid topographic data. However, averaging elevations could also introduce loss of information to the final 2D model schematics. In some recent research, new grid adjustment methods could minimise such losses and improve accuracy of simulation results with reducing computational cost (Chen et al.,

2012a). Either traditional or new grid adjustment method still need a proper quality of topographic data for representing key component of urban features. Adopting MSV approach could be capable of providing that topographic data.

9.7 3Di for enhancing 2D models

Whenever 2D flood models were simulated, almost all modellers are wishful to get their completed results as fast as possible. Typically, many conventional 2D flood models consume a lot of computational costs especially when a large number of regular grids are computing. Nowadays, some advanced software of 2D flood models are capable of simulating such many grids faster than such conventional 2D flood models.

Amongst these new 2D flood models these days, the 3Di software could be count as an outstanding 2D flood model. The 3Di solver (Stelling, 2012) in this software is fully packed with four essential numerical methods, i.e. the sub-grid method (Casulli, 2009), the bathymetry friction, the finite-volume staggered grid method (Stelling & Duinmeijer, 2003), and quad-tree techniques (Wang et al., 2004). In Casulli & Stelling (2011), sub-grid method is used for computing water levels on conventional grid cells, but advances in this technique are that they can calculate conveyance values only for fine grid cells (Fig. 9-10). In this way, flow solvers remain fast while all detailed elevation changes are felt by the flow.

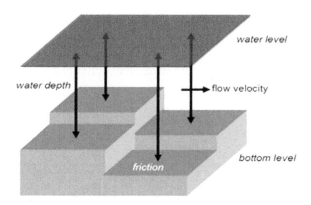

Fig. 9-10. An example of the basic drivers in
sub-grid method (source by 3Di)

This new 3Di software with sub-grid method can handle a large number of
computational grids of high-resolution topographic data. With this software, 3Di
developers claimed that modellers can achieve faster runtimes compared to
conventional flood models. In this respect, simulating floods either for rural or
complex urban areas can be customised (Fig. 9-11), when applying sub-grid
methods of the 3Di software.

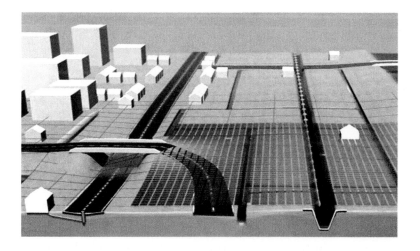

Fig. 9-11. An example of a schematic map of the rural area (on the right) and
the complex urban area (on the left) using sub-grid method (source by 3Di)

Whenever or wherever more detailed key components of the complex drainage capabilities (Fig. 9-12a) are seriously concerned, large number of computational grids with high-resolution details could reveal real benefits and push this powerful 3Di software up to their limit capabilities.

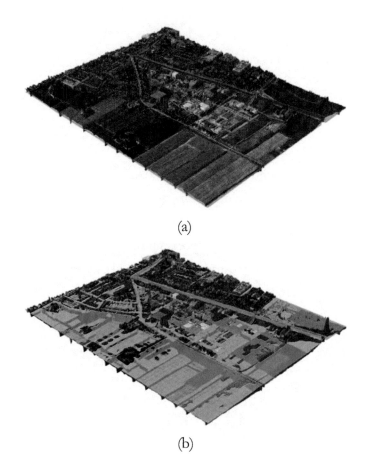

(a)

(b)

Fig. 9-12. (a) A fine detail topographic data of a city with the normal water level.
(b) High floodwaters simulated in fine detail (source by 3Di)

Despite sampling, translating, and preparing raster-base topographic data as sophisticated input, directly importing point cloud topographic data is yet another great feature that can be easily applied in this 3Di software. Directly importing point cloud data can foreshorten some steps of creating 2D model schematics. Such point cloud data can be used later for representing simulation results with better visualisations (Fig. 9-12b). With this high-resolution point cloud data, analysing more details of urban areas concerning buildings, pavements, parks, and retentions in the level of urban feature details is feasible.

In 3Di software, the simulation results can be represented in a virtual 3D environment also, their model setups and simulations results are centrally stored in 3Di cloud services. Technically, this value information will never get lost. Creating emergency storages, raising dyke heights locally, or breaching certain segments can be done on the fly. Existing model setups can be upgraded and modified remotely in nearly real-time. Modifying model setups, re-calculating models, and visualising simulation results can be handled easily at almost everywhere over a web browser whenever the Internet accesses are available. Almost real-time information can be accessed through an interactive web portal and realistic 3D-animations for a clearer picture of complex flood situations.

From trial experiences, this 3Di software could be counted as a new versatile water management instrument based on detailed hydraulic computations. The fast-computational claims from 3Di developers also sounded promising for now. And for the future, we should explore more for the full functional capabilities of 3Di software of handling a large number of computational grids for simulating complex-flood situations.

9.8 High-performance computers for minimising computational costs

In conventional 2D flood models, many coarse computational grids can represent only a few indications of the whole complex flood simulation. With the fine computational grids, they should bring more key indications than those coarse grids. Technically, when 2D flood models were computed, a large number computational grids are commonly spent a lot of computational costs. Higher performance computers can play a significant role for shortening computing time for each simulation. Without doubt, a large number computational grids of complex flood simulation will take enormous time or event not can be done without helping of such high-performance computers. For speeding up 2D model simulations, enabling these parallel computing techniques: Message Passing Interface (MPI), Open Multi Processing (Open-MP), and General-Purpose computing on Graphics Processing Units (GPGPU) in 2D model software should achieve full capabilities of CPUs and GPUs of parallel computers

Although some 2D flood-model software these days can fully gain a lot benefits of multi-processing capabilities for running their simulations, other software cannot reach that full capabilities even on a stand-alone computer. Apart from software limitations, we would further focus on the hardware capabilities and their limitations. Generally, it may be better to have more than one stand-alone computer in order to run many simulations in parallel. Whenever one can get many computers with exactly the same specs of such stand-alone computers as they want, setting those computers may always be time-consuming. Operating and maintaining such computers separately need intensive cares in order to perform them well without interruptions or any evidentially fail down by losing power or processor overheated.

From Chapter 7, around 400,000 computational cells at 20 m grid resolution 2D models of Ayutthaya case study were used as an example for comparing some capacities of parallel processing. This 2D model used MIKE 21™ were running on a standard laptop equipped with two physical cores of Intel® i5 CPU at 2.30 GHz, 4 GB of RAM and it spent ~18 running hours for a single simulation result. When using a more advanced laptop equipped with four physical cores Intel® i7

CPU at 2.20 GHz, 16 GB of RAM, and 2 GB Video RAM embedded in NVIDIA® GeForce™ GTX580M graphics cards, it spent ~14 running hours for the same simulation setups. However, these laptops still cannot compete with high-performance cloud-computing system of SURFsara service (free of charges for students and staffs and available for some Dutch academies). Due to physical hardware of SURFsara contains a lot more CPU and/or GPU processors than the stand-alone computer. Thus, they can perform much faster.

In this SURFsara service, their thirty nodes of each cloud-computing server equipped with thirty-two processors of which have 16 physical cores Intel® Xeon® E5-2698 v3 at 2.7 GHz, and 256 GB of RAM. For this research, our virtual machines (VMs) on SURFsara were running with 40 cores and 64 GB of RAM. Although running MIKE 21™ simulation can compatible with 8 cores, our VMs on SURFsara only spent ~8 running hours for the same simulation setups.

Moreover, having a good care of SURFsara services, some operating and maintenance issues could be minimised. Even though setting those virtual machines (VMs) on SURFsara services may take times for beginners, once the first VM can be properly created, duplicating a bunch of such VMs is easy and reliable. As almost the same capabilities of other cloud services, modifying model setups, re-calculating models, and visualising simulation results can also be handled easily in each VM at almost everywhere and anytime over a web browser whenever the Internet accesses are available.

CHAPTER 10
Conclusions and recommendations

Modelling urban flood dynamics and its complex processes require proper representation of complex topographies. Structure from Motion (SfM) techniques bring new opportunities of achieving better quality topographic data by using different viewpoints. This dissertation explores insights into the capabilities of fusing SfM point cloud data by using different-source views as input for enhancing 2D urban flood modelling. The benefit of merging side-view SfM data with conventional top-view LiDAR data was explored using the novel multi-source views (MSV) approach. Making use of these new MSV topographic input data led to promising results to enhance 2D model schematizations. Simulation results appear to represent more realistic flood dynamics, especially in complex cities. Implementing the new MSV data approach was explored for two case studies: (i) the 2003 flood event of Kuala Lumpur, Malaysia (Chapter 6) and (ii) the 2011 flood event of Ayutthaya Island, Thailand (Chapters 7 and 8). In this chapter, all findings related to the research questions (Section 1.3) are presented (Sections 10.1 to 10.5, resp), while recommendations are given in Section 10.6.

10.1 Limitations of using conventional top-view LiDAR data

Three main limitations of using conventional top-view Light Detection and Ranging (LiDAR) data are:

(i) Airborne LiDAR surveys have difficulties in observing some key components of urban features: vertical structures (e.g. building walls, alleys, openings, and watermarks), and low-level structures (e.g. pathways and kerbs), since airborne LiDAR systems pulse signals target surfaces only vertically. Most urban features are exposed to the sky (e.g. building rooftops and ground terrains) and can easily be observed by these airborne LiDAR systems. However, other concealed or covered urban features are hard to observe by LiDAR, because they are concealed and sometimes relatively small.

(ii) Utilising conventional top-view LiDAR data as input (even in high resolution and high accuracy) does not always result in better flood model schematizations. In this research, we used conventional top-view LiDAR data as input for creating 2D schematics from different types of DEMs: LiDAR Digital Surface Model (LiDAR-DSM; Section 5.3.1 for a hypothetical case; Section 6.2.1 for a real case study); LiDAR Digital Terrain Model (LiDAR-DTM; Section 5.3.2 for a hypothetical case; Section 6.2.2 and 7.2.2 for real case studies); LiDAR Digital Building Model with terrains (LiDAR-DBM+; Section 5.3.3 for a hypothetical case; Section 7.2.2 for a real case study).

Applying these 2D schematics for urban flood models seem to be practical for most situations. However, some key components of urban features were still missing and are not incorporated in conventional top-view LiDAR data. In fact, some features allow floodwaters flow through, while others obstruct or diver floodwater flows differently. Such (missing) features may have significant impact on complex flood dynamics in cities (Razafison et al., 2012). (iii) Conventional top-view LiDAR data have long been used in various applications. When accessibility to high-resolution LiDAR data is lacking, acquiring such top-view LiDAR data often requires a lot of investments.

10.2 Benefits of using SfM technique for creating topographic data

When obtaining topographic data from conventional top-view surveys, some key components of complex urban features can be overlooked. Obtaining data from different viewpoints can bring a better analysis of complex urban features. Emerging of digital photos and Structure from Motion (SfM) techniques is capable of achieving high resolutions and high accuracies of topographic data from different viewpoints.

In this research, a new side-view Structure from Motion (SfM) surveys were introduced (Chapter 4). These side-view SfM surveys are capable to acquire some missing urban features (e.g. alleys, pathways, low-level structures, and openings). In contrast, such missing key features can hardly be detected from conventional top-view LiDAR surveys.

Since advances in computer vision and photogrammetry techniques have dramatically improved, the SfM techniques can provide appropriate quality of 3D point cloud data in sub-metre resolutions. Unlike other advanced topographic data acquisitions (e.g. LiDAR system), the low-cost digital cameras, camcorders, or even mobile phone cameras can be adequately used to achieve digital photo resolutions. Such relatively small cameras can be easily mounted on bicycle, vehicles, or even used as handheld devices, which are easy to operate in the field.

By using standard digital photo and video scenes in combination with SfM techniques, one can create 3D point cloud data from overlapping 2D photos. When the overlapping photos have been selected, the key aspect of the SfM technique can then be used to determine the matching 3D positions of each key point in the overlapping 2D photos. While repeatability means of same key points were detected in different images. Then descriptor extracted essential information of each identified key point. When such descriptive key points were assigned, they were ready for matching with its neighbour overlapping photos. A track linking specific key points in a set of overlapping photos can then be established. Following the matching step, the sparse bundle adjustment system was used to estimate camera pose. It also used to improve the accuracy, minimise projection errors from camera tracking, and refined the camera positions in order to extract a low-density or 'sparse' SfM point cloud data. Finally, triangulation is used to estimate the raw 3D point cloud data.

The findings showed that the new side-view SfM surveys can be adequately used to obtain essential information of complex urban features. Distinct key components of urban features, i.e. arches, openings, sky-train tracks, high trees, alleys, low-level structures, and watermarks were explored and highlighted in Chapter 5, 6, and 7, resp.

10.3 3D point cloud data can be fused for constructing proper elevation maps

This present work explores the capabilities of fusing 3D point cloud data obtained from different views for enhancing 2D urban flood modelling. Some key components of urban features may still be misrepresented or hardly be observed when obtaining topographic data from single viewpoint.

Whenever such topographic data are merged, these merged topographic data from different viewpoints should bring benefits from their both different viewpoints. Extracting key components of urban features from the conventional top-view LiDAR point cloud data applied a straightforward approach to extract and classify terrains, buildings, and trees (Chapter 3), but only terrain point cloud and building point cloud were mainly used. When the new side-view SfM point cloud was registered in the same geo-referencing system (e.g. EPSG 4326: WGS 84) of conventional top-view LiDAR data, the new side-view SfM data can be analysed and merged with conventional top-view LiDAR point cloud data to create novel Multi-Source Views (MSV) point cloud data. The edited 3D point cloud of MSV data were ready for creating elevation maps when all irritant high features, i.e. overarching structures and high trees were properly eliminated (Chapter 5).

Making use of the new side-view SfM point cloud also showed that it can be used for flood watermark extraction (Chapter 7). Flood peaks are often difficult to find either manually or from gauges distributed on land. It may even be difficult to get fully recorded time-series data for analysing flood peaks. Whenever such observed times-series data are inappropriate to analyse, SfM data can be used for flood watermarks detection. Typically, conventional land surveys are used for measuring peak elevation of watermarks. Trash lines created watermarks on some urban structures (e.g. walls and poles) often last long, for several months or years. In post-flood analyses, verifying flood peaks is crucial and commonly used as benchmark for evaluating simulation results (Gee et al., 1990; Bates et al., 1992).

From Ayutthaya case study, existing flood watermarks were observed two years after the 2011 flood event for Ayutthaya. Two mobile units using side-view SfM surveys were used for data acquisition, speeding up the time-consuming and intensive work of conventional land surveying. Side-view SfM topographic data were then embedded into new MSV topographic data for flood peak extraction. The new MSV data showed good agreement with conventional data observed by land surveys. Making use of MSV data provides opportunities for better analysing historical flood magnitudes and their elevation benchmarks, which looks promising for verifying simulation results (Chapter 7).

10.4 3D point cloud data can be used for enhancing 2D flood models

Even though the new MSV point cloud data are better for representing complex urban features, urban flood modelling mainly aims to provide an approximation of reality. The essence of modelling is that it represents and replicates essentials, leaving out irrelevant details. Very high details of edited MSV point cloud data were simplified for creating raster-based DEMs (using ArcGIS® commercial software), which were then incorporated into 2D model schematics (using MIKE Zero™ commercial software). Choices of such 2D schematic adjustments are crucially depending on spatial and temporal resolutions of each flood situation (Chapter 5).

Examples of merging different-source view topographic data were applied to two case studies: (i) For the Kuala Lumpur case study, new MSV point cloud data were created by merging ~0.20 m resolution side-view SfM data with ~0.50 m resolution conventional top-view LiDAR data and all edited point cloud were then rasterized to created DEMs at 1 m grid resolution (Chapter 6); (ii) For the Ayutthaya case study, new MSV data were created by merging ~0.20 m resolution side-view SfM data with ~0.63 m resolution top-view LiDAR data and all edited point cloud were then rasterized to created DEMs at 5 m, 10 m, and 20 m grid resolutions (Chapter 7).

In urban features analysis, merging side-view SfM data with conventional top-view LiDAR data into new MSV data showed that most key components of urban features could be extracted. Moreover, when applying MSV digital elevation model (DEM) as input, some key components of complex urban features (e.g. retention walls under trees, pathways under arches, and kerbs under sky-train tracks), can also be taken into account for enhancing 2D model schematics (examples in Section 6.2.3 and 7.3.2). Some algorithms were developed to distinguish between walls and openings (algorithms in Chapter 4; examples in Chapter 7).

There are several ways of identifying and determining such complex openings using side-view surveys (e.g. field surveys, street-view photo archives, and human interpretations), which are applicable for enhancing 2D model schematics. In this research, side-views from SfM point cloud data were used to create MSV-DEM for enhancing 2D model schematics. We also compared simulation results using conventional top-view LiDAR DEMs with the new MSV-DEM as input for urban flood models for the two case studies (Kuala Lumpur and Ayutthaya).

The findings showed that simulation results using conventional top-view (i) LiDAR-DSM as input have the least flood inundation areas and results represented many dry areas. Because (i) LiDAR-DSM contains many urban features including buildings, high trees, and overarching structures, which behave as obstacles in 2D model schematics. When applying conventional top-view (iii) LiDAR-DBM+ and new (iv) MSV-DEM as input, their simulation results showed more inundation areas close to reality. The results represented almost the same size of flood inundations. However, some small missing inundation areas were now more revealed and better represented when applying (iv) MSV-DEM as input. Comparing simulated results of using (iii) LiDAR-DBM+ and (iv) MSV-DEM did not reflect a significant difference (Chapter 7).

In this research, adjusting manning's M values was evaluated for roughness sensitivities on both case studies. We incorporated different Manning's M roughness values for 1D river networks, which were valuable for adjusting simulation results. We found that when uniformly applying Manning's M at 40 on rivers, simulation results showed good agreement with observed measurements for both case studies (Section 6.3 and 7.6). When uniformly applying three different Manning's M at 20, 30, and 40 for floodplain of Ayutthaya case study, we found that simulated results only showed insignificant flood dynamic changes at the initial stage. However, adopting Manning's M at 30 on Ayutthaya's floodplain, simulation results showed good agreement with observed measurements.

In dynamic flood situations, rapid changes occur. When simulating such dynamic features, increasing the time step and embedding more spatial details of obstacles could better represent flood velocities and flood flow routes. Grid modifications and equivalent roughness adjustments can simulate effects of openings to enhance 2D model schematics. These spatial details can play a significant role in representing complex flood dynamics and for analysing complex flood evacuation routes.

Floodwater depths and flood inundation patterns may just have minor changes during most of the flood event. When simulating the end result, floods become almost static (stagnant) and adopting simple terrain topographic data can represent decent urban-flood inundation maps, which could give reasonable information for further analysing flood depth damages.

Modifying complex key components in 2D model schematics is less sensitive for urban-flood mapping. Many researches appear to represent adequate solutions based on filtering conventional top-view LiDAR data for enhancing urban-flood mapping (Schubert et al., 2008; Tsubaki & Fujita, 2010; Abdullah et al., 2012a; Abdullah et al., 2012b; Vojinović et al., 2013a).

In this research, conventional LiDAR-DTM was applied as input for urban-flood mapping in the Ayutthaya case study. When comparing simulated results of using different DEMs, the findings showed that using LiDAR-DTM as input represented fastest flood time evolutions for fully covered inundations. Because filtering methods remove most of the urban obstacles, replacing them with new flat elevations from surrounding ground values, the almost completely flat surfaces in LiDAR-DTM let floodwaters flow freely on their 2D model schematics. Even though such simulated results cannot represent flow routes properly, their flood map results of floodwater depths and flood inundation areas can provide reasonable information to help developing local and regional flood-protection measures for local stakeholders (Chapter 8).

10.5 Enhanced computer-based environments can help developing flood-protection measures

From Ayutthaya case study (Chapter 8), it can be seen that applying computer-based environments can help in simulating and visualising different flood-protection measures. In this way, the best locations for gates and pumping stations at Makamreang Canal and Thor Canal were indicated as a first priority. A second measure was related to create a multifunctional landscape by upgrading and restoring ancient canals and ponds. A third measure was to increase dyke heights up to $0.3 - 0.5$ m from existing elevations by using the U-Thong Road as a ring road dyke. With new alternative flood pathways, using existing roads together with ancient canal and pond networks (Section 8.4.4) can be applied for conveying excess floodwaters to allocated storage areas. Holistic flood risk assessment approaches for this case study gained benefits from computer-based environments, which is also emphasised in Vojinović et al. (2016).

Again, the new MSV approach can help in indicating and extracting some essential features. For some key components (e.g. alleys, retention walls in Kuala Lumpur case study also roads, kerbs, low-level structures, and some hidden pathways in the Ayutthaya case study), the advances in MSV data also helped in developing and enhancing 2D model schematizations for both case studies. New flood pathways revealed some capabilities of using road drainage capacities for critical infrastructure planning and for business-interruption loss estimations. When key components in model schematics were appropriately analysed, proposing alternative flood-prevention and flood-adaptation measures can also be improved for evaluating flood damages and losses (Evans et al., 2008). Thereafter, new alternative measures will be promising to help developing better flood protection measures.

Nowadays, advances in urban flood simulation need higher resolution topographic data to be used as input. Key components of urban features could bring better understanding to complex floods and in turn also provide valuable information for flood-protection measure alternatives. Proper analyses in flood model schematics can provide decent simulation results with reasonable explanations, leading to more practical flood-protection measures.

High-performance computational simulation models can play a significant role for shortening simulation times for high-resolution urban-flood modelling. For a stand-alone computer, simulation models can perform adequately for simple cases. Using cloud computing services (e.g. SURFsara services), real powerful parallel computing can far-better show increased performances of advanced models. Since cloud computers commonly contain a lot more CPU and/or GPU processors than any stand-alone computer, as a result, they can perform much faster. For advanced simulation models available in the market, the 3Di model can be seen as one of the recent advances in urban flood simulation and visualisation.

10.6 Recommendations

Automated algorithms toward a better processing for multi-views topographic data

Multi-views surveys can bring new benefits from both top and side viewpoints for urban analyses and urban flood management applications, as shown in two case studies (Chapters 6 and 7). Even though some algorithms gave examples for indicating key components of urban features, merging multi-source views topographic data, and translating such data to 2D schematic cells, further automatic algorithms further need to be developed to better automate the entire process. Developing such further algorithms should be capable of indicating, extracting, merging, and translating such key components for enhancing 2D schematics automatically with a little help from experts.

Advances in small-format surveying platforms

Conventionally, manned airborne platforms are aeroplanes, helicopters, hot-air balloons, and sailplanes, which are commonly operated by experienced pilots. Despite requiring pilot-trained skills, most unmanned vehicles allow pilots operating their aircrafts from the ground (with less experiences or even never have a pilot-trained skill). Thankful for new technology developments that made surveying tools become smaller and lighter, most small-format aircrafts these days are capable of caring many surveying tools. Some outstanding unmanned airborne platforms highlighted that they can obtain high-resolution topographic data from different viewpoints: from top-view using UAV aerial-based surveys (Section 9.2), from side-view using MMS land-based surveys (Section 9.3), and from underwater-view using USV aquatic-based surveys (Section 9.4).

A new era of unmanned surveying platforms should have high portability, rapid setup, easy to use, and minimal requirements of well-trained operators. They should be utilised in situations that would be impractical, risky, or impossible to be operated in conventional ways. They should also have a capability of worries-free functions (e.g. return to the base station, self-geolocating identifications, and/or safety parachute for aerial vehicles) in order to rescue such surveying platforms including saving civilians from any accident for every surveying operation.

Advances in small-format digital cameras and SfM technique

During the last decades, low-cost digital cameras seem to be the most popular use for not only daily life usages but also surveying purposes. Nowadays, several new digital cameras tend to show that they could provide more photo resolutions. Further developments of such cameras should be capable of instant usage at almost everywhere (even underwater) at any time (low-light situation capability) when one wants to observe their surroundings. The future SfM applications on mobile (phone) cameras should be developed and make use of further development in SfM techniques should provide more qualities and accuracies of 3D point cloud data.

Advances in numerical flood simulations

New 3Di engines provide a new possibility for running very detailed hydraulic simulation and their computations are extremely fast and accurate for complex urban-flood predictions. Advances in 3Di software can generate an immediate picture of effects of measures. Their user interface enables decision makers and civilians to visualise impacts of the proposed measures in various climate scenarios. Outstanding capabilities of 3Di software should provide more benefits of using MSV point cloud directly for setting up model schematics and representations.

Advances in parallel computing

In this research, the real power of parallel computing can be highlighted when using the SURFsara cloud-computing services. Once a virtual machine is created, duplicating a bunch of such VMs is easy and fast and maintenance issues are minimised. In SURFsara services, physical hardware of cloud computers contains a lot of CPU and/or GPU processors far better than any stand-alone computer. As a result, they can perform much faster. Without doubt, running our simulation will take enormous time (if) without using this high-performance cloud-computing services. Further developments of grid and cloud-computing units, equipped with CPUs and GPUs are underway at HAII/MOST, Thailand. Developing such computing capabilities should be further provided for academic and science communities, in somehow at almost the same as SURFsara computing cloud services provided for years in the Netherlands.

References

3Di. 3Di Watermanagement, Deltares, <www.3di.nu>. Accessed Dec 2015.

Abbott M.B. and Ionescu F., 1967. On The Numerical Computation Of Nearly Horizontal Flows. Journal of Hydraulic Research, 5 (2): 97-117. doi: 10.1080/00221686709500195.

Abbott M.B. Computational hydraulics: Elementary of the theory of free surface flow London: Pitman Publishing Ltd.; 1980.

Abbott M.B. and Vojinović Z., 2009. Applications of numerical modelling in hydroinformatics. Journal of Hydroinformatics, 11 (3-4): 308-19. doi: 10.2166/hydro.2009.051.

Abbott M.B. and Vojinović Z., 2010a. Realising social justice in the water sector: 1. Journal of Hydroinformatics, 12 (1): 97-117. doi: 10.2166/hydro.2010.065.

Abbott M.B. and Vojinović Z., 2010b. Realising social justice in the water sector: 2. Journal of Hydroinformatics, 12 (2): 225-39. doi: 10.2166/hydro.2009.065.

Abbott M.B. and Vojinović Z., 2013. Towards a hydroinformatics for China. Journal of Hydroinformatics, 15 (4): 1189-202. doi: 10.2166/hydro.2012.178.

Abbott M.B. and Vojinović Z., 2014. Towards a hydroinformatics praxis in the service of social justice. Journal of Hydroinformatics, 16 (2): 516-30. doi: 10.2166/hydro.2013.198.

Abdullah A.F., Vojinović Z., Price R.K., and Aziz N.A.A., 2012a. Improved methodology for processing raw LiDAR data to support urban flood modelling - accounting for elevated roads and bridges. Journal of Hydroinformatics, 14 (2): 253-69. doi: 10.2166/hydro.2011.009.

Abdullah A.F., Vojinović Z., Price R.K., and Aziz N.A.A., 2012b. A methodology for processing raw LiDAR data to support urban flood modelling framework. Journal of Hydroinformatics, 14 (1): 75-92. doi: 10.2166/hydro.2011.089.

Aber J.S., Marzolff I., and Ries J. Small-Format Aerial Photography: Principles, Techniques and Geoscience Applications: Elsevier; 2010.

Ackermann F. The accuracy of digital terrain models. In. 37th Photogrammetric Week, University of Stuttgart, Germany 1979.

ADB. Flood Risk Assessment Study for The Historic City of Ayutthaya, Thailand: Final Report. Bangkok: Asian Development Bank (ADB); 2015.

AHN 2. LiDAR of Delft, The Netherlands. Upper left coner 85067.56 mE 447473.68 mN, Lower right coner 85395.16 mE 791.2 mN, WGS84 zone 31U; 2014.

Aktaruzzaman M. and Schmidt T. Detailed digital surface model (DSM) generation and automatic object dection to facilitate modelling of urban flooding. In. ISPRS Workshop, Hannover 2009.

All4Desktop. Small River Creek, All For Desktop: Licensed under CC-BY2.0, <all4desktop.com/4245787-river.html>. Accessed Oct 2016.

Alonzo M., Bookhagen B., and Roberts D.A., 2014. Urban tree species mapping using hyperspectral and lidar data fusion. Remote Sensing of Environment, 148: 70-83. doi: 10.1016/j.rse.2014.03.018.

Aronica G., Bates P.D., and Horritt M.S., 2002. Assessing the uncertainty in distributed model predictions using observed binary pattern information within GLUE. Hydrological Processes, 16 (10): 2001-16. doi: 10.1002/hyp.398.

Awrangjeb M., Zhang C., and Fraser C.S., 2013. Automatic extraction of building roofs using LIDAR data and multispectral imagery. ISPRS Journal of Photogrammetry and Remote Sensing, 83: 1-18. doi: 10.1016/j.isprsjprs.2013.05.006.

Axelsson P., 1999. Processing of laser scanner data—algorithms and applications. ISPRS Journal of Photogrammetry and Remote Sensing, 54 (2–3): 138-47. doi: 10.1016/S0924-2716(99)00008-8.

Axelsson P., 2000. DEM generation from laser scanner data using adaptive TIN models. International Archives of Photogrammetry and Remote Sensing, 33 (B4/1; PART 4): 111-18.

Ballesteros Cánovas J.A., Eguibar M., Bodoque J.M., Díez-Herrero A., Stoffel M., and Gutiérrez-Pérez I., 2011. Estimating flash flood discharge in an ungauged mountain catchment with 2D hydraulic models and dendrogeomorphic palaeostage indicators. Hydrological Processes, 25 (6): 970-79. doi: 10.1002/hyp.7888.

Balmforth D., Digman C.J., Butler D., and Shaffer P. Research Framework - The implementation of Integrated Urban Drainage. London: DEFRA; 2006.

Baltsavias E.P., 1999. A comparison between photogrammetry and laser scanning. ISPRS Journal of Photogrammetry and Remote Sensing, 54 (2–3): 83-94. doi: 10.1016/S0924-2716(99)00014-3.

Bates P.D., Anderson M.G., Baird L., Walling D.E., and Simm D., 1992. Modelling floodplain flows using a two-dimensional finite element model. Earth Surface Processes and Landforms, 17 (6): 575-88. doi: 10.1002/esp.3290170604.

Bates P.D., Wilson M.D., Horritt M.S., Mason D.C., Holden N., and Currie A., 2006. Reach scale floodplain inundation dynamics observed using airborne synthetic aperture radar imagery: Data analysis and modelling. Journal of Hydrology, 328 (1–2): 306-18. doi: 10.1016/j.jhydrol.2005.12.028.

Bellocchio F., Borghese N.A., Ferrari S., and Piuri V. 3D Surface Reconstruction: Multi-scale hierarchical approaches Dordrecht: Springer; 2013.

Berehulak D. People make for the safety of a bridge above flooded streets in Pathum Thani, The Guardian, 2011, <www.theguardian.com/world/gallery/2011/oct/23/thailand-flooding-threatens-bangkok-gallery>. Accessed Nov 2015.

Berry J.K. Map analysis: understanding spatial patterns and relationships: GeoTec Media; 2007.

Beven K., 2000. On the future of distributed modelling in hydrology. Hydrological Processes, 14 (16-17): 3183-84. doi: 10.1002/1099-1085(200011/12)14:16/17 <3183::AID-HYP404>3.0.CO;2-K.

BOE. Weekly Epidemiological Surveillance Report. Bangkok: Bureau of Epidemiology (BOE), National Trustworthy and Competent Authority in Epideriological Surveillance and Investigation; 2011.

Boonya-aroonnet S. Applications of the innovative modelling of urban surface flooding in the UK case studies. 11th International Conference on Urban Drainage, Edinburgh, UK; August 31-September 5, 2008, 2008.

Boonya-aroonnet S. Applications of the innovative modelling of urban surface flooding in the UK case studies. 11th International conference on urban drainage, Edinburgh, UK2010.

Boonya-aroonnet S., Tantianuparp P., Weesakul S., and Chitradon R. Technology Assisted Flood Management. In. PAWEES 2015 International Conference on Climate Change and Water & Environment Management in Monsoon Asia, Bangkok, Thailand: International Society of Paddy and Water Environment Engineering (PAWEES); Nov 27-29, 2015.

BOT. Thailand flood 2011: Impact and recovery from business survey. Bangkok: Bank of Thailand (BOT); 2012.

Brivio P.A., Colombo R., Maggi M., and Tomasoni R., 2002. Integration of remote sensing data and GIS for accurate mapping of flooded areas. International Journal of Remote Sensing, 23 (3): 429-41. doi: 10.1080/01431160010014729.

Bull L.J., Kirkby M.J., Shannon J., and Hooke J.M., 2000. The impact of rainstorms on floods in ephemeral channels in southeast Spain. CATENA, 38 (3): 191-209. doi: 10.1016/S0341-8162(99)00071-5.

Casulli V., 2009. A high-resolution wetting and drying algorithm for free-surface hydrodynamics. International Journal for Numerical Methods in Fluids, 60 (4): 391-408. doi: 10.1002/fld.1896.

Casulli V. and Stelling G.S., 2011. Semi-implicit subgrid modelling of three-dimensional free-surface flows. International Journal for Numerical Methods in Fluids, 67 (4): 441-49. doi: 10.1002/fld.2361.

Catarinella M. Amsterdam summer university: Connecting cultures since 1990, Amsterdam Summer University (AMSU), <amsu.eu/euroxperience/>. Accessed Mar 2015.

Chaudhry M. Applied Hydraulic Transients New York: Van Nostrand Reinhold Co.; 1979.

Chen A.S., Hsu M.H., Teng W.H., Huang C.J., Yeh S.H., and Lien W.Y., 2006. Establishing the database of inundation potential in Taiwan. Nat Hazards, 37: 107-32. doi: 10.1007/s11069-005-4659-7.

Chen A.S., Evans B., Djordjević S., and Savić D.A., 2012a. A coarse-grid approach to representing building blockage effects in 2D urban flood modelling. J Hydrol, 426–427: 1-11. doi: 10.1016/j.jhydrol.2012.01.007.

Chen A.S., Evans B., Djordjević S., and Savić D.A., 2012b. Multi-layered coarse grid modelling in 2D urban flood simulations. J Hydrol, 470-471: 1-11. doi: 10.1016/j.jhydrol.2012.06.022.

Chen Z., Xu B., and Gao B., 2015. Assessing visual green effects of individual urban trees using airborne Lidar data. Science of The Total Environment, 536: 232-44. doi: 10.1016/j.scitotenv.2015.06.142.

Chezy A. Formule pour trouver la vitesse de l'eau conduit dan une rigole donnée. Dossier 847 (MS 1915) of the manuscript collection of the École Nationaldes Ponts et Chaussées, Paris. Reproduced in: Mouret, G.(1921). Antoine Chézy: histoire d'une formule d'hydraulique. In. Annales des Ponts et Chaussées 1776.

Chow V.T. Open-channel hydraulics New York McGraw-Hill; 1959.

Colwell R.N. History and place of photographic interpretation. In: Philipson W.R. (Ed.). Manual of photographic interpretation, 2nd ed. Bethesda, Maryland: AmericanSociety for Photogrammetry and Remote Sensing 1997. pp. 3-47.

Connell R., Painter D., and Beffa C., 2001. Two-Dimensional Flood Plain Flow.II: Model Validation. Journal of Hydrologic Engineering, 6 (5): 406-15. doi: 10.1061/(ASCE) 1084-0699(2001)6:5(406).

Cunge J.A., Holly F.M., and Verwey A. Practical aspects of computational river hydraulics: monographs and surveys in water resources engineering London: Pitman Publishing Ltd; 1980.

Cunge J.A., 2003. Of data and models. Journal of Hydroinformatics, 5: 75-98.

Dai F. and Lu M., 2010. Assessing the Accuracy of Applying Photogrammetry to Take Geometric Measurements on Building Products. Journal of Construction Engineering and Management, 136 (2): 242-50. doi: 10.1061/(ASCE)CO.1943-7862.0000114.

Dai F., Rashidi A., Brilakis I., and Vela P., 2013. Comparison of Image-Based and Time-of-Flight-Based Technologies for Three-Dimensional Reconstruction of Infrastructure. Journal of Construction Engineering and Management, 139 (1): 69-79. doi: 10.1061/(ASCE)CO.1943-7862.0000565.

Dai F., Feng Y., and Hough R., 2014. Photogrammetric error sources and impacts on modeling and surveying in construction engineering applications. Visualization in Engineering, 2 (1): 1-14. doi: 10.1186/2213-7459-2-2.

De Saint-Venant A.B., 1871. Théorie du mouvement non permanent des eaux, avec application aux crues des rivières et à l'introduction des marées dans leurs lits. Comptes Rendus des séances de l'Académie des Sciences, 73: 237-40.

De Sherbinin A., Schiller A., and Pulsipher A., 2007. The vulnerability of global cities to climate hazards. Environment and Urbanization, 19 (1): 39-64. doi: 10.1177/0956247807076725.

Denlinger R.P., O'Connell D.R.H., and House P.K. Robust Determination of Stage and Discharge: An Example from an Extreme Flood on the Verde River, Arizona. Ancient Floods, Modern Hazards: American Geophysical Union; 2002. pp. 127-46.

DHI. Klang River-basin environment improvement and flood mitigation project (Stormwater Management and Road Tunnel - SMART). Final report. Kuala Lumpur: Department of Irrigation and Drainage; 2004.

DHI. MIKE 11: A Modelling System for Rivers and Channels - Reference Manual Horsolm, Denmark: MIKE by DHI; 2016a.

DHI. MIKE 21 Flow Model: Hydrodynamic Module - User Guide Horsolm, Denmark: MIKE by DHI; 2016b.

DHI. MIKE FLOOD: 1D-2D Modelling - User Manual Horsolm, Denmark: MIKE by DHI; 2016c.

Djordjević S., Prodanović D., and Walters G., 2004. Simulation of transcritical flow in pipe/channel networks. J Hydraul Eng, 130: 1167-78. doi: 10.1061/(ASCE)0733-9429(2004)130:12(1167).

Drury S.A. Image Interpretation in Geology London: Allen & Unwin; 1987.

Ducke B., Score D., and Reeves J., 2011. Multiview 3D reconstruction of the archaeological site at Weymouth from image series. Computers & Graphics, 35 (2): 375-82. doi: 10.1016/j.cag.2011.01.006.

Durbin P.A. and Iaccarino G., 2002. An Approach to Local Refinement of Structured Grids. Journal of Computational Physics, 181 (2): 639-53. doi: 10.1006/jcph.2002.7147.

Dutta D., Herath S., and Musiake K., 2000. Flood inundation simulation in a river basin using a physically based distributed hydrologic model. Hydrological Processes, 14 (3): 497-519. doi: 10.1002/(SICI)1099-1085(20000228)14:3<497::AID-HYP951>3.0.CO;2-U.

Edelman A., Arias T.A., and Smith S.T., 1998. The Geometry of Algorithms with Orthogonality Constraints. SIAM Journal on Matrix Analysis and Applications, 20 (2): 303-53. doi: 10.1137/S0895479895290954.

Elbers H. The top-view scene of Rijksmuseum, Amsterdam, The Netherlands, Flickr: licensed under CC-BY2.0, 2013, <www.flickr.com/photos/hanselpedia/10164795726/in/photostream>. Accessed Mar 2015.

Elmqvist M., Jungert E., Lantz F., and Persson A. Terrain modelling and analysis using laser scanner data. International Archives of Photogrammetric & Remote Sensing, XXXIV-3 (4): 219-26.

EM-DAT. The OFDA/CRED International Disaster Database. Brussels, Belgium: Université Catholique de Louvain; 2014.

Erdody T.L. and Moskal L.M., 2010. Fusion of LiDAR and imagery for estimating forest canopy fuels. Remote Sensing of Environment, 114 (4): 725-37. doi: 10.1016/j.rse.2009.11.002.

Evans B., Simm J.D., Thorne C.R., Arnell N.W., Hess T.M., and Lane S.N. The Pitt Review: An Update of the Foresight Future Flooding 2004 Qualitative Risk Analysis. Cabinet Office, London; 2008.

Evans E.P., Wicks J.M., Whitlow C.D., and Ramsbottom D.M. The evolution of a river modelling system. In. Proceedings of the Institution of Civil Engineers - Water Management 2007.

Fairfield J. and Leymarie P., 1991. Drainage networks from grid digital elevation models. Water Resources Research, 27 (5): 709-17.

Ferraz A., Mallet C., and Chehata N., 2016. Large-scale road detection in forested mountainous areas using airborne topographic lidar data. ISPRS Journal of Photogrammetry and Remote Sensing, 112: 23-36. doi: 10.1016/j.isprsjprs. 2015.12.002.

Fewtrell T.J., Bates P.D., Horritt M., and Hunter N.M., 2008. Evaluating the effect of scale in flood inundation modelling in urban environments. Hydrological Processes, 22 (26): 5107-18. doi: 10.1002/hyp.7148.

Fewtrell T.J., Duncan A., Sampson C.C., Neal J.C., and Bates P.D., 2011. Benchmarking urban flood models of varying complexity and scale using high resolution terrestrial LiDAR data. Phys Chem Earth Pt A/B/C, 36 (7-8): 281-91. doi: 10.1016/j.pce.2010.12.011.

Filippova O. and Hänel D., 1998. Grid Refinement for Lattice-BGK Models. Journal of Computational Physics, 147 (1): 219-28. doi: 10.1006/jcph.1998.6089.

Fischler M.A. and Bolles R.C., 1981. Random sample consensus: a paradigm for model fitting with applications to image analysis and automated cartography. Commun ACM, 24 (6): 381-95. doi: 10.1145/358669.358692.

Fraser C.S., 1997. Digital camera self-calibration. ISPRS Journal of Photogrammetry and Remote Sensing, 52 (4): 149-59.

Furukawa Y. and Ponce J., 2010. Accurate, Dense, and Robust Multiview Stereopsis. IEEE Transactions on Pattern Analysis and Machine Intelligence, 32 (8): 1362-76. doi: 10.1109/TPAMI.2009.161.

Gallegos H.A., Schubert J.E., and Sanders B.F., 2009. Two-dimensional, high-resolution modeling of urban dam-break flooding: A case study of Baldwin Hills, California. Advances in Water Resources, 32 (8): 1323-35. doi: 10.1016/j.advwatres. 2009.05.008.

García-Pintado J., Neal J.C., Mason D.C., Dance S.L., and Bates P.D., 2013. Scheduling satellite-based SAR acquisition for sequential assimilation of water level observations into flood modelling. Journal of Hydrology, 495 (0): 252-66. doi: 10.1016/j.jhydrol.2013.03.050.

Gaume E. and Borga M., 2008. Post-flood field investigations in upland catchments after major flash floods: proposal of a methodology and illustrations. Journal of Flood Risk Management, 1 (4): 175-89. doi: 10.1111/j.1753-318X.2008.00023.x.

Gee D.M., Anderson M.G., and Baird L., 1990. Large-scale floodplain modelling. Earth Surface Processes and Landforms, 15 (6): 513-23. doi: 10.1002/esp.3290150604.

GFDRR. Thai flood 2011: Rapid assessment for resilient recovery and reconstruction planning. Bangkok: Global Facility for Disaster Reduction and Recovery (GFDRR), International Bank for Reconstruction and Development; 2012.

Goesele M., Snavely N., Curless B., Hoppe H., and Seitz S. Multi-view stereo for community photo collections. IEEE 11th ICCV, Rio de Janeiro, Brazil2007. doi: 10.1109/ICCV.2007.4408933.

Golparvar-Fard M., Bohn J., Teizer J., Savarese S., and Peña-Mora F., 2011. Evaluation of image-based modeling and laser scanning accuracy for emerging automated performance monitoring techniques. Automation in Construction, 20 (8): 1143-55. doi: 10.1016/j.autcon.2011.04.016.

Goodell C. Quasi Two-Dimensional Modeling in HEC-RAS, The RAS Solution: The Place for HEC-RAS Modelers, 2013, <hecrasmodel.blogspot.nl/2013/03/quasi-two-dimensional-modeling-in-hec.html>. Accessed May 2015.

Google Earth™ 7.1.5.1557. The lowland area of Ayutthaya Island in dry season. 14°20'44.35" N, 100°32'58.00" E, eye alt 2.5 km: Aerodata International Surveys; 2013.

Google Earth™ 7.1.5.1557. The lowland area of Ayutthaya Island in the begining of wet season. 14°20'44.35" N, 100°32'58.00" E, eye alt 2.5 km: Aerodata International Surveys; 2014.

Google Earth™ 7.1.5.1557. The complex urban scene of Delft, The Netherlands. 52°00'47" N, 4°22'12" E, eye alt 500 m: Aerodata International Surveys; 2016a.

Google Earth™ 7.1.5.1557. Top-view image satellite of Rijksmuseum, Amsterdam, The Netherlands. 52°21'38.87" N, 4°53'06.57" E, eye alt 1.0 km: DigitalGlobe; 2016b.

Google Earth™ 7.1.5.1557. Complex urban fabric of Delft City, The Netherlands. 52°00'34.2"N 4°22'00.7"E, eye alt 500 m: Aerodata International Surveys; 2017.

Google Maps™. The city map of Kuala Lumpur, Malaysia. 3° 8'54.48" N, 101°41'44.09" W, eye alt 1 km: Google; 2012a.

Google Maps™. The city map of Phra Nakhon Si Ayutthaya, Thailand. 14°21'11.46" N, 100°34'8.28" E, eye alt 10km: Google; 2012b.

Gordon P. and Charles K.T. Introduction to Laser Ranging, Profiling, and Scanning. Topographic Laser Ranging and Scanning: CRC Press; 2008. pp. 1-28.

Goyer G. and Watson R., 1963. The laser and its application to meteorology. Bulletin of the American Meteorological Society, 44 (9): 564-75.

Graves B. Remote control, small-format aerial photography: The Easy Star way. US: Earth Science Department, Emporia State University, Kansas; 2007.

Green C., 2004. The evaluation of vulnerability to flooding. Disaster Prevention and Management: An International Journal, 13 (4): 323-29. doi: 10.1108/09653560 410556546.

Gross H. and Thoennessen U. Extraction of lines from laser point clouds. In. Symposium of ISPRS Commission III: Photogrammetric Computer Vision PCV06 International Archives of Photogrammetry, Remote Sensing and Spatial Information Sciences 2006.

Haala N., Brenner C., and Anders K.H., 1998. 3D urban GIS from laser altimeter and 2D map data. International Archives of Photogrammetric & Remote Sensing, 32 (Part 3/1): 339-46.

Habib A., Kersting A., Ruifang Z., Al-Durgham M., Kim C., and Lee D., 2008. LiDAR strip adjustment using conjugate linear features in overlapping strips. Int Arch Photogramm Rem Sens and Spat Inf Sci, 37: 385-90.

Hai P.T., Magome J., Yorozuya A., Inomata H., Fukami K., and Takeuchi K., 2010. Large-scale flooding analysis in the suburbs of Tokyo Metropolis caused by levee breach of the Tone River using a 2D hydrodynamic model. Water Sci Technol, 62: 1859-64. doi: 10.2166/wst.2010.381.

HAII. ThaiWeather.Net: Storm tracking for Thailand, Hydro and Agro Informatics Institute, Ministry of Science and Technology (HAII/MOST), 2012, <www.thaiwater.net/TyphoonTracking/storm/storm.html>. Accessed Jan 2015.

Haile A.T. and Rientjes T.H. Effects of LiDAR DEM resolution in flood modelling: A model sensitivity study for the city of Tegucigalpa, Honduras. In: Vosselman G.& Brenner C., (Eds.). 36th International Conference Society for Photogrammetry and Remote Sensing: ISPRS Workshop 'Laser Scanning', Volume XXXVI-3/W19, 2005, WG III/3-4 V/3, Enschede, The Netherlands; September 12-14, 2005, 2005a.

Haile A.T. and Rientjes T.H. Effects of LiDAR DEM resolution in flood modelling: a model sensitivity study for the city of Tegucigalpa, Honduras. 36th ISPRS Workshop 'Laser Scanning 2005', Enschede, The Netherlands2005b.

Ham W.E. and Curtis N.M.J. Common minerals, rocks, and fossils of Oklahoma: Oklahoma Geological Survey; 1960.

Han K.-Y., Lee J.-T., and Park J.-H., 1998. Flood inundation analysis resulting from Levee-break. Journal of Hydraulic Research, 36 (5): 747-59. doi: 10.1080/00221689 809498600.

Han W., Zhao S., Feng X., and Chen L., 2014. Extraction of multilayer vegetation coverage using airborne LiDAR discrete points with intensity information in urban areas: A case study in Nanjing City, China. International Journal of Applied Earth Observation and Geoinformation, 30: 56-64. doi: 10.1016/j.jag.2014.01.016.

Hervouet J. and Janin J. Finite element algorithms for modelling flood propagation. In. Modelling of Flood Propagation over Initially Dry Areas, Milan, Italy 1994.

Hervouet J. and Van-Haren L. Recent advances in numerical methods for fluid flow. In: Anderson M.G., Walling D.E.& Bates P.D. (Eds.). Floodplain Processes. Chiehester: Wiley; 1996. pp. 183-14.

Hervouet J.M., 2000. A high resolution 2-D dam-break model using parallelization. Hydrological Processes, 14 (13): 2211-30. doi: 10.1002/1099-1085(200009)14:13 <2211::AID-HYP24>3.0.CO;2-8.

Hesselink A.W., Stelling G.S., Kwadijk J.C.J., and Middelkoop H., 2003. Inundation of a Dutch river polder, sensitivity analysis of a physically based inundation model using historic data. Water Resources Research, 39 (9): 1-17. doi: 10.1029/2002WR001334.

Hestholm S. and Ruud B., 1994. 2D finite-difference elastic wave modelling including surface topography. Geophys Prospect, 42: 371-90. doi: 10.1111/j.1365-2478.1994.tb00216.x.

Hooke J.M. and Mant J.M., 2000. Geomorphological impacts of a flood event on ephemeral channels in SE Spain. Geomorphology, 34 (3–4): 163-80. doi: 10.1016/S0169-555X(00)00005-2.

Horritt M.S., 2000. Calibration of a two-dimensional finite element flood flow model using satellite radar imagery. Water Resources Research, 36 (11): 3279-91. doi: 10.1029/2000WR900206.

Horritt M.S. and Bates P.D., 2001. Effects of spatial resolution on a raster based model of flood flow. Journal of Hydrology, 253: 239-49. doi: 10.1016/S0022-1694(01)00490-5.

Hsu M.H., Chen S.H., and Chang T.J., 2000a. Inundation simulation for urban drainage basin with storm sewer system. Journal of Hydrology, 234 (1–2): 21-37. doi: 10.1016/S0022-1694(00)00237-7.

Hsu M.H., Chen S.H., and Chang T.J., 2000b. Inundation simulation for urban drainage basin with storm sewer system. J Hydrol, 234: 21-37. doi: 10.1016/S0022-1694(00)00237-7.

Hunter N.M., Bates P.D., Neelz S., Pender G., Villanueva I., Wright N.G., Liang D., Falconer R.A., Lin B., and Waller S., 2008a. Benchmarking 2D hydraulic models for urban flooding. P Ice-water Manage, 161: 13-30. doi: 10.1680/wama.2008.161.1.13.

Hunter N.M., Bates P.D., Neelz S., Pender G., Villanueva I., Wright N.G., Liang D., Falconer R.A., Lin B., and Waller S. Benchmarking 2D hydraulic models for urban flooding. In. Proceedings of the Institution of Civil Enginees - Water Management 161; February, 2008, 2008b. doi: 10.1680/wama.2008.161.1.13.

Jadidi H., Ravanshadnia M., Hosseinalipour M., and Rahmani F., 2015. A step-by-step construction site photography procedure to enhance the efficiency of as-built data visualization: a case study. Visualization in Engineering, 3 (1): 1-12. doi: 10.1186/s40327-014-0016-9.

Jancosek M. and Pajdla T., 2011. Multi-View Reconstruction Preserving Weakly-Supported Surfaces. 2011 Ieee Conference on Computer Vision and Pattern Recognition (Cvpr).

Jensen J.R. Remote sensing of the environment : An earth resource perspective. Prentice Hall series in geographic information science Upper Saddle River, New Jersey: Prentice Hall; 2000.

Jha A., Lamond J., Bloch R., Bhattacharya N., Lopez A., Papachristodoulou N., Bird A., Proverbs D., Davies J., and Barker R. Five feet high and rising: citites and flooding in the 21st century. Washington, D.C.: World Bank; 2011.

JICA, NESDB, RID, and DWR. Executive Summary of the Flood Management Plan for the Chao Phraya River Basin in the Kingdom of Thailand. Office of National Economic and Social Development Board (NESDB), Royal Irrigation Department, Ministry of Agriculture and Cooperatives (RID/MOAC), Department of Water Resources, Ministy of Natural Resources and Environment (DWR/MNRE), and Japan International Cooperation Agency (JICA); 2013.

Joe T. and Nigel W. Aspects of numerical grid generation - Current science and art. 11th Applied Aerodynamics Conference: American Institute of Aeronautics and Astronautics; 1993.

Johnson A. Surveying at CUED: GPS at Waddlescairn, Department of Engineering, University of Cambridge, 2004, <www2.eng.cam.ac.uk/~alj3/Survey/4M9pics/31%20GPS%20at%20Waddlescairn%201.JPG>. Accessed Dec 2015.

Jongman B., Ward P.J., and Aerts J.C.J.H., 2012. Global exposure to river and coastal flooding: Long term trends and changes. Global Environmental Change, 22: 823-35. doi: 10.1016/j.gloenvcha.2012.07.004.

Kabolizade M., Ebadi H., and Ahmadi S., 2010. An improved snake model for automatic extraction of buildings from urban aerial images and LiDAR data. Computers, Environment and Urban Systems, 34 (5): 435-41. doi: 10.1016/j.compenvurbsys.2010.04.006.

Keerakamolchai W. Towards a Framework for Multifunctional Flood Detention Facilities Design in a Mixed Land Use Area: The Case of Ayutthaya World Heritage Site, Thailand (MSc thesis thesis), AIT, Pathumthani and UNESCO-IHE, Delft; 2014.

Kilian J., Haala N., and Englich M., 1996. Capture and evaluation of airborne laser scanner data. International Archives of Photogrammetry and Remote Sensing, 31: 383-88.

Knotters M. and Bierkens M.F.P., 2001. Predicting water table depths in space and time using a regionalised time series model. Geoderma, 103 (1–2): 51-77. doi: 10.1016/S0016-7061(01)00069-6.

Lane S.N., James T.D., Pritchard H., and Saunders M., 2003. Photogrammetric and laser altimetric reconstruction of water levels for extreme flood event analysis. The Photogrammetric Record, 18 (104): 293-307. doi: 10.1046/j.0031-868X.2003.00022.x.

Leandro J., Schumann A., and Pfister A., 2016. A step towards considering the spatial heterogeneity of urban key features in urban hydrology flood modelling. Journal of Hydrology, 535: 356-65. doi: 10.1016/j.jhydrol.2016.01.060.

Lezenby G. Canal of Delft, Flickr: licensed under CC-BY2.0, 2013, <www.flickr.com/photos/65089906@N00/9103173908>. Accessed Mar 2015.

Lhomme J., Bouvier C., Mignot E., and Paquier A., 2006. One-dimensional GIS-based model compared with a two-dimensional model in urban floods simulation. Water Science & Technology, 54 (6-7): 83-91.

Li Y., Yong B., Wu H., An R., and Xu H., 2015. Road detection from airborne LiDAR point clouds adaptive for variability of intensity data. Optik - International Journal for Light and Electron Optics, 126 (23): 4292-98. doi: 10.1016/j.ijleo.2015.08.137.

Liang D., Falconer R., and Lin B. Linking one- and two-dimensional models for free surface flows. In. Proceedings of the Institution of Civil Engineers - Water Management 2007.

Lin B., Wicks J.M., Falconer R.A., and Adams K. Integrating 1D and 2D hydrodynamic models for flood simulation. In. Proceedings of the Institution of Civil Engineers - Water Management 2006.

Lowe D.G. Object recognition from local scale-invariant features. In. Computer Vision, 1999 The Proceedings of the Seventh IEEE International Conference on 1999. doi: 10.1109/ICCV.1999.790410.

Lowe D.G., 2004. Distinctive image features from scale-invariant keypoints. Int J Comput Vision, 60: 91-110. doi: 10.1023/ B:VISI.0000029664.99615.94.

Malin D. and Light D.l. Aerial Photography A2 - Peres, Michael R. The Focal Encyclopedia of Photography (Fourth Edition). Boston: Focal Press; 2007. pp. 501-04.

Manning R., Griffith J.P., Pigot T., and Vernon-Harcourt L.F. On the flow of water in open channels and pipes; 1890.

Marchi L., Borga M., Preciso E., Sangati M., Gaume E., Bain V., Delrieu G., Bonnifait L., and Pogačnik N., 2009. Comprehensive post-event survey of a flash flood in Western Slovenia: observation strategy and lessons learned. Hydrological Processes, 23 (26): 3761-70. doi: 10.1002/hyp.7542.

Mark O., Weesakul S., Apirumanekul C., Aroonnet S.B., and Djordjević S., 2004. Potential and limitations of 1D modelling of urban flooding. Journal of Hydrology, 299 (3–4): 284-99. doi: 10.1016/j.jhydrol.2004.08.014.

Marks K. and Bates P.D., 2000. Integration of high-resolution topographic data with floodplain flow models. Hydrol Process, 14: 2109-22. doi: 10.1002/1099-1085(20000815/30)14:11/12<2109::AID-HYP58>3.0.CO;2-1.

Martz L.W. and Garbrecht J., 1992. Numerical definition of drainage network and subcatchment areas from Digital Elevation Models. Computers & Geosciences, 18 (6): 747-61. doi: 10.1016/0098-3004(92)90007-E.

Mason D.C., Schumann G.J.P., Neal J.C., Garcia-Pintado J., and Bates P.D., 2012. Automatic near real-time selection of flood water levels from high resolution Synthetic Aperture Radar images for assimilation into hydraulic models: A case study. Remote Sensing of Environment, 124 (0): 705-16. doi: 10.1016/j.rse.2012.06.017.

Maune D.F. Digital elevation model technologies and applications: the DEM users manual: American Society for Photogrammetry and Remote Sensing; 2007.

McCowan A.D., Rasmussen E.B., and Berg P. Improving the performance of a two-dimensional hydraulic model for floodplain applications. Conference on Hydraulics in Civil Engineering, Hobart, Australia; Nov 28-30, 2001.

McCoy A.P., Golparvar-Fard M., Rigby E.T., and 2014. Reducing Barriers to Remote Project Planning: Comparison of Low-Tech Site Capture Approaches and Image-Based 3D Reconstruction. Journal of Architectural Engineering, 20 (1): 05013002. doi: 10.1061/(ASCE)AE.1943-5568.0000118.

Meesuk V., Vojinović Z., and Mynett A.E. Merging top-view LiDAR data with street-view SfM data to enhance urban flood simulation. 11th International Conference on Hydroinformatics - HIC 2014, New York, USA2014.

Mitasova H., Hardin E., Starek M.J., Harmon R.S., Overton M.F., Hengl T., Evans I., Wilson J., and Gould M. Landscape dynamics from LiDAR data time series. Geomorphometry, CA, USA2011.

Mongus D., Lukač N., and Žalik B., 2014. Ground and building extraction from LiDAR data based on differential morphological profiles and locally fitted surfaces. ISPRS Journal of Photogrammetry and Remote Sensing, 93: 145-56. doi: 10.1016/j.isprsjprs.2013.12.002.

Morales-Hernández M., Petaccia G., Brufau P., and García-Navarro P., 2016. Conservative 1D–2D coupled numerical strategies applied to river flooding: The Tiber (Rome). Applied Mathematical Modelling, 40 (3): 2087-105. doi: 10.1016/j.apm.2015.08.016.

Mynett A.E. and Vojinović Z., 2009. Hydroinformatics in multi-colours—part red: urban flood and disaster management. Journal of Hydroinformatics, 11 (3-4): 166-80. doi: 10.2166/hydro.2009.027.

Neal J.C., Bates P.D., Fewtrell T.J., Hunter N.M., Wilson M.D., and Horritt M.S., 2009. Distributed whole city water level measurements from the Carlisle 2005 urban flood event and comparison with hydraulic model simulations. J Hydrol, 368: 42-55. doi: 10.1016/j.jhydrol.2009.01.026.

Néelz S. and Pender G., 2007. Sub-grid scale parameterisation of 2D hydrodynamic models of inundation in the urban area. Acta Geophysica, 55 (1): 65-72. doi: 10.2478/s11600-006-0039-2.

Neubronner J. Die Brieftaubenphotographie und ihre Bedeutung für die Kriegskunst, als Doppelsport, für die Wissenschaft und im Dienste der Presse (in German) Dresden: Wilhelm Baensch; 1909.

Nocedal J. and Wright S.J. Numerical optimization. Springer Science; 1999.

Osterman M. The Technical Evolution of Photography in the 19th Century A2 - Peres, Michael R. The Focal Encyclopedia of Photography (Fourth Edition). Boston: Focal Press; 2007. pp. 27-36.

Overton I.C., 2005. Modelling floodplain inundation on a regulated river: integrating GIS, remote sensing and hydrological models. River Research and Applications, 21 (9): 991-1001. doi: 10.1002/rra.867.

Panya Consultants. The feasibility study of the east diversion channel of Chao Phraya river basin. Thai Ministry of Agriculture and Cooperatives' Royal Irrigation Department (RID); 2012.

Pappenberger F., Beven K.J., Hunter N.M., Bates P.D., Gouweleeuw B.T., Thielen J., and De Roo A.P.J., 2005. Cascading model uncertainty from medium range weather forecasts (10 days) through a rainfall-runoff model to flood inundation predictions within the European Flood Forecasting System (EFFS). Hydrology and Earth System Sciences, 9 (4): 381-93.

PCL. Point Cloud Library (PCL): Module registration, Point Cloud Library, <docs. pointclouds.org/trunk/group__registration.html>. Accessed Mar 2015.

Peaceman D.W. and Rachford H.H., 1955. The numerical solution of parabolic and elliptic differential equations. Journal of the Society for Industrial and Applied Mathematics, 3 (1): 28-41.

Phungwong N., 2012. The interpretation of Europeans settlements (Portuguese, Dutch and French) on Chao Phraya River during Ayutthaya Era. Veridian E-Journal, 5 (2): 91-119.

Preissmann A. Propagation des intumescences dans les canaux et rivieres. In. First Congress French Assoc for Computation 1961.

Price R.K. and Vojinović Z. Urban Hydroinformatics: Data, Models, and Decision Support for Integrated Urban Water Management London: IWA publishing; 2011.

PsychaSec. Amsterdam Alley, Flicrk: Licensed under CC-BY2.0, 2013, <www.flickr.com/ photos/psychasec/9095485940>. Accessed Mar 2016.

Puente I., González-Jorge H., Martínez-Sánchez J., and Arias P., 2013. Review of mobile mapping and surveying technologies. Measurement, 46 (7): 2127-45. doi: 10.1016/j.measurement.2013.03.006.

Razafison U., Cordier S., Delestre O., Darboux F., Lucas C., and James F., 2012. A shallow water model for the numerical simulation of overland flow on surfaces with ridges and furrows. Eur J Mechanics B-fluids, 31: 44-52. doi: 10.1016/j.euromechflu. 2011.07.002.

Reeves M. Modelling of pressure pipes (including forcemains) in InfoWorks ICM and CS, Innovyze, 2013, <blog.innovyze.com/2013/02/18/modelling-of-pressure-pipes-including-forcemains-in-infoworks-icm-and-cs/>. Accessed May 2016.

Remondino F. and El-Hakim S., 2006. Image-based 3D Modelling: A Review. The Photogrammetric Record, 21 (115): 269-91. doi: 10.1111/j.1477-9730.2006. 00383.x.

Rodionov B.N., Isavnina I.V., Avdeev Y.F., Blagov V.D., Dorofeev A.S., Dunaev B.S., Ziman Y.L., Kiselev V.V., Krasikov V.A., Lebedev O.N., Mikhailovskii A.B., Tishchenko A.P., Nepoklonov B.V., Samoilov V.K., Truskov F.M., Chesnokov Y.M., and Fivenskii Y.I., 1971. New data on the Moon's figure and relief based on results from the reduction of Zond-6 photographs. Cosmic Research, 9: 410-17.

Roisri A. Flooded Historical Temple in Ayutthaya, Thailand, 123RF Limited, 2011, <https://nl.123rf.com/profile_arunroisri>. Accessed Dec 2016.

Romanowicz R. and Beven K., 2003. Estimation of flood inundation probabilities as conditioned on event inundation maps. Water Resources Research, 39 (3): 4-1 to 12. doi: 10.1029/2001WR001056.

Rosenqvist Å., Forsberg B.R., Pimentel T., Rauste Y.A., and Richey J.E., 2002. The use of spaceborne radar data to model inundation patterns and trace gas emissions in the central Amazon floodplain. International Journal of Remote Sensing, 23 (7): 1303-28. doi: 10.1080/01431160110092911.

Rottensteiner F. and Briese C., 2002. A new method for building extraction in urban areas from high-resolution LIDAR data. International Archives of Photogrammetry Remote Sensing and Spatial Information Sciences, 34 (3/A): 295-301.

Rychkov I., Brasington J., and Vericat D., 2012. Computational and methodological aspects of terrestrial surface analysis based on point clouds. Computers & Geosciences, 42: 64-70. doi: 10.1016/j.cageo.2012.02.011.

Sampson C.C., Fewtrell T.J., Duncan A., Shaad K., Horritt M.S., and Bates P.D., 2012. Use of terrestrial laser scanning data to drive decimetric resolution urban inundation models. Advances in Water Resources, 41: 1-17. doi: 10.1016/j.advwatres.2012.02.010.

Samuels P.G. Cross section location in one-dimensional models. In. International Conference on river 1990.

Sandercock P.J. and Hooke J.M., 2010. Assessment of vegetation effects on hydraulics and of feedbacks on plant survival and zonation in ephemeral channels. Hydrological Processes, 24 (6): 695-713. doi: 10.1002/hyp.7508.

Schenk T. and Csathó B., 2002. Fusion of LIDAR data and aerial imagery for a more complete surface description. International Archives of Photogrammetry Remote Sensing and Spatial Information Sciences, 34 (3/A): 310-17.

Schubert J.E. and Sanders B.F., 2012. Building treatments for urban flood inundation models and implications for predictive skill and modeling efficiency. Adv Water Resour, 41: 49-64. doi: 10.1016/j.advwatres.2012.02.012.

Seyoum S.D., Vojinović Z., Price R.K., Weesakul S., and 2012. Coupled 1D and Noninertia 2D Flood Inundation Model for Simulation of Urban Flooding. Journal of Hydraulic Engineering, 138 (1): 23-34. doi: 10.1061/(ASCE)HY.1943-7900.0000485.

Simoes N., Leitão J.P., Maksimović, Č, Sá Marques A., Pina R., and 2010. Sensitivity analysis of surface runoff generation in urban flood forecasting. Water Science & Technology, 61 (10): 2595-601. doi: 10.2166/wst.2010.178.

Singh R., Arya D.S., Taxak A.K., and Vojinović Z., 2016. Potential Impact of Climate Change on Rainfall Intensity-Duration-Frequency Curves in Roorkee, India. Water Resources Management: 1-14. doi: 10.1007/s11269-016-1441-4.

Smith M.W., 2014. Roughness in the Earth Sciences. Earth-Science Reviews, 136: 202-25. doi: 10.1016/j.earscirev.2014.05.016.

Smith R.A.E., Bates P.D., and Hayes C., 2012. Evaluation of a coastal flood inundation model using hard and soft data. Environ Modell Softw, 30: 35-46. doi: 10.1016/j.envsoft.2011.11.008.

Snavely N., Seitz S.M., and Szeliski R. Photo tourism: exploring photo collections in 3D. In. ACM transactions on graphics (TOG) 2006. doi: 10.1145/1179352.1141964.

Snavely N., Seitz S.M., and Szeliski R., 2007. Modeling the World from Internet Photo Collections. International Journal of Computer Vision, 80 (2): 189-210. doi: 10.1007/s11263-007-0107-3.

Sohn G. and Dowman I., 2007. Data fusion of high-resolution satellite imagery and LiDAR data for automatic building extraction. ISPRS Journal of Photogrammetry and Remote Sensing, 62 (1): 43-63. doi: 10.1016/j.isprsjprs.2007.01.001.

Song J., Wu J., and Jiang Y., 2015. Extraction and reconstruction of curved surface buildings by contour clustering using airborne LiDAR data. Optik - International Journal for Light and Electron Optics, 126 (5): 513-21. doi: 10.1016/j.ijleo.2015.01.011.

Stelling G., Kernkamp H., and Laguzzi M. Delft flooding system: a powerful tool for inundation assessment based upon a positive flow simulation. HIC, Copenhagen, Denmark1998.

Stelling G.S. and Duinmeijer S.P.A., 2003. A staggered conservative scheme for every Froude number in rapidly varied shallow water flows. International Journal for Numerical Methods in Fluids, 43 (12): 1329-54. doi: 10.1002/fld.537.

Stelling G.S., 2012. Quadtree flood simulations with sub-grid digital elevation models. Proceedings of the Institution of Civil Engineers - Water Management, 165 (10): 567-80. doi: 10.1680/wama.12.00018.

Struys J.J. Illustrations de Les voyages de Jean Struys en Moscovie: Bibliothèque nationale de France; 1718.

Szeliski R. and Kang S.B., 1994. Recovering 3D Shape and Motion from Image Streams Using Nonlinear Least Squares. Journal of Visual Communication and Image Representation, 5 (1): 10-28. doi: 10.1006/jvci.1994.1002.

Townsend P.A. and Foster J.R., 2002. A synthetic aperture radar–based model to assess historical changes in lowland floodplain hydroperiod. Water Resources Research, 38 (7): 20-1-20-10. doi: 10.1029/2001WR001046.

Toyoda Y., Taniguchi H., Siyanee H., and Pongpisit H. Values of Ayutthaya Historical Park: Promoting willingness to pay for flood protection. In. Disaster Mitigation of Urban Cultural Heritage Proceedings 2012.

Ukon T., Shigeta N., Watanabe M., Shiraishi H., and Uotani M. Correction methods for dropping of simulated water level utilising Preissmann and MOUSE slot models. In. 11th International conference on urban drainage 2008.

UN. Strengthening the Global Partnership for Development in a Time of Crisis. MDG Task Force Report. Geneva: United Nations (UN); 2009.

UN News Centre. As flood disaster worsens in Thailand, UN steps up relief efforts, United Nations, 2011, <www.un.org/apps/news/story.asp?NewsID=40233#. VdwklvmqpBd>. Accessed Aug 2015.

van Dijk E., van der Meulen J., Kluck J., and Straatman J., 2014. Comparing modelling techniques for analysing urban pluvial flooding. Water Science & Technology, 69 (2): 305-11. doi: 10.2166/wst.2013.699.

Vincent L., 2007. Taking online maps down to street level. Computer, 40 (12): 118-20.

Virginia Living Museum. Floodwater through the doors of main museum building, Howell Creative Group, <thevlm.org/im-starting-to-feel-a-little-like-han-solo-here/>. Accessed May 2016.

Vojinović Z., Solomatine D., and Price R.K., 2006. Dynamic least-cost optimisation of wastewater system remedial works requirements. Water Science and Technology, 54 (6-7): 467-75. doi: 10.2166/wst.2006.574.

Vojinović Z., 2007. A complementary modelling approach to manage uncertainty of computationally expensive models. Water Science and Technology, 56 (8): 1-9. doi: 10.2166/wst.2007.599.

Vojinović Z. and Van Teeffelen J., 2007. An integrated stormwater management approach for small islands in tropical climates. Urban Water Journal, 4 (3): 211-31. doi: 10.1080/15730620701464190.

Vojinović Z. and Tutulic D., 2009. On the use of 1D and coupled 1D-2D modelling approaches for assessment of flood damage in urban areas. Urban Water J, 6: 183-99. doi: 10.1080/15730620802566877.

Vojinović Z., Seyoum S.D., Mwalwaka J.M., and Price R.K., 2011. Effects of model schematisation, geometry and parameter values on urban flood modelling. Water Sci Technol, 63: 462-67. doi: 10.2166/wst.2011.244.

Vojinović Z. and Abbott M.B. Flood risk and social justice: From quantitative to qualitative flood risk assessment and mitigation: IWA Publishing (International Water Assoc); 2012.

Vojinović Z., Abebe Y.A., Ranasinghe R., Vacher A., Martens P., Mandl D.J., Frye S.W., van Ettinger E., and de Zeeuw R., 2013. A machine learning approach for estimation of shallow water depths from optical satellite images and sonar measurements. Journal of Hydroinformatics, 15 (4): 1408-24. doi: 10.2166/hydro.2013.234.

Vojinović Z., Hammond M., Golub D., Hirunsalee S., Weesakul S., Meesuk V., Medina N., Sanchez A., Kumara S., and Abbott M., 2016. Holistic approach to flood risk assessment in areas with cultural heritage: a practical application in Ayutthaya, Thailand. Natural Hazards, 81 (1): 589-616. doi: 10.1007/s11069-015-2098-7.

Wang J.P., Borthwick A.G.L., and Taylor R.E., 2004. Finite-volume-type VOF method on dynamically adaptive quadtree grids. International Journal for Numerical Methods in Fluids, 45 (5): 485-508. doi: 10.1002/fld.712.

Wang X., Cao Z., Pender G., and Neelz S., 2010. Numerical modelling of flood flows over irregular topography. Proceedings of the Institution of Civil Engineers - Water Management, 163 (5): 255-65. doi: 10.1680/wama.2010.163.5.255.

Wehr A. and Lohr U., 1999. Airborne laser scanning—an introduction and overview. ISPRS Journal of Photogrammetry and Remote Sensing, 54 (2–3): 68-82. doi: 10.1016/S0924-2716(99)00011-8.

Wehr A. LIDAR: Airborne and terrestrial sensors. In: Li Z., Chen J.& Baltsavlas E. (Eds.). Advances in Photogrammetry, Remote Sensing and Spatial Information Sciences: 2008 ISPRS Congress Book. Leiden: CRC Press; 2008. pp. 73-84.

Werner M., Blazkova S., and Petr J., 2005. Spatially distributed observations in constraining inundation modelling uncertainties. Hydrological Processes, 19 (16): 3081-96. doi: 10.1002/hyp.5833.

Westoby M.J., Brasington J., Glasser N.F., Hambrey M.J., and Reynolds J.M., 2012. 'Structure-from-Motion' photogrammetry: A low-cost, effective tool for geoscience applications. Geomorphology, 179: 300-14. doi: 10.1016/j.geomorph.2012.08.021.

Wildi E. Master Composition Guide for Digital Photographers: Amherst Media, Inc.; 2006.

Wright N.G. Introduction to numerical methods for fluid flow. In: Bates P.D., Lane S.N.& Ferguson R.I. (Eds.). Computational Fluid Dynamics: Applications in Environmental Hydraulics. Chichester: Wiley; 2005.

Wu C.C. SiftGPU: A GPU implementation of Scale Invaraint Feature Transform (SIFT), University of North Carolina at Chapel Hill, 2007, <cs.unc.edu/~ccwu/siftgpu>. Accessed Dec 2011.

Wu C.C. VisualSFM: A Visual Structure from Motion System, University of Washington, 2011, <ccwu.me/vsfm/>. Accessed Dec 2011.

Wu C.C., Agarwal S., Curless B., and Seitz S.M. Multicore Bundle Adjustment. IEEE Conference on Computer Vision and Pattern Recognition, CO, USA2011. doi: 10.1109/CVPR.2011.5995552.

Xiao J., Fang T., Tan P., Zhao P., Ofek E., and Quan L., 2008. Image-based façade modeling. ACM Trans Graph, 27 (5): 1-10. doi: 10.1145/1409060.1409114.

Yan W.Y., Shaker A., and El-Ashmawy N., 2015. Urban land cover classification using airborne LiDAR data: A review. Remote Sensing of Environment, 158: 295-310. doi: 10.1016/j.rse.2014.11.001.

Yao W., Hinz S., and Stilla U., 2010. Automatic vehicle extraction from airborne LiDAR data of urban areas aided by geodesic morphology. Pattern Recognition Letters, 31 (10): 1100-08. doi: 10.1016/j.patrec.2010.02.006.

Yu D. and Lane S.N., 2006. Urban fluvial flood modelling using a two-dimensional diffusion-wave treatment, part 1: mesh resolution effects. Hydrological Processes, 20 (7): 1541-65. doi: 10.1002/hyp.5935.

Yuan X., Xie Z., and Liang M., 2008. Spatiotemporal prediction of shallow water table depths in continental China. Water Resources Research, 44 (4): W04414. doi: 10.1029/2006WR005453.

Zahorcak M. Evolution of the photographic lens in the 19th century. In: Peres M.R. (Ed.). Focal encyclopedia of photography: Digital imaging, theory and applications, history, and science, 4th ed. Amsterdam: Elsevier; 2007.

Zevenbergen C., Cashman A., Evelpidou N., Pasche E., Garvin S., and Ashley R. Urban flood management Leiden: CRC Press; 2010.

Zhang K., Yan J., and Chen S.C., 2006. Automatic Construction of Building Footprints From Airborne LIDAR Data. IEEE Transactions on Geoscience and Remote Sensing, 44 (9): 2523-33. doi: 10.1109/TGRS.2006.874137.

Zhou G., Song C., Simmers J., and Cheng P., 2004. Urban 3D GIS From LiDAR and digital aerial images. Computers & Geosciences, 30 (4): 345-53. doi: 10.1016/j.cageo.2003.08.012.

About the author

Vorawit Meesuk was born on 25 May 1977 in Bangkok, Thailand. In 1999, he obtained his B.Sc. in Computer Science from the Rajamangala Institute of Technology, Pathum Thani, Thailand. In 2003, he received his M.Sc. in Remote Sensing Technology and GIS from the Khon Kaen University, Thailand, where he was later  appointed as Associate Researcher at the Ground Water Centre. He continued until mid 2004 when he took up the position of Associate Researcher at the Hydro and Agro Informatics Institute (HAII), Ministry of Science and Technology (MOST), Bangkok, Thailand. In April 2011, he started his PhD research at UNESCO-IHE/TU Delft, the Netherlands. His PhD research focused on the topic of fusing point cloud data for enhancing 2D urban flood modelling. Besides doing research, he organised a workshop in Ayutthaya, in a joint effort with ADB, UNESCO-BANGKOK, in 2014. He also guided M.Sc. students and was co-lecturer for GIS and urban-flood modelling classes at UNESCO-IHE until 2016.

Currently, he works at HAII/MOST as Head of Observation and Telemetry Section, whose responsibilities are providing and maintaining telemetry systems for monitoring weather conditions and water-level changes in Thailand and neighbouring countries. Until now, his team has carried out telemetry observations in over 700 stations.

He is married to Sutchaleo. Their most enjoyable activity is hiking deep into mountains far away from complex urban cities.

Journal Papers

Meesuk, V., Vojinović, Z., Mynett, A.E., and Abdullah, A.F. (2015). Urban Flood Modelling Combining Top-view LiDAR data with Ground-view SfM Observations. Advances in Water Resources, 75, 105-117. doi: 10.1016/j.advwatres.2014.11.008.

Vojinović, Z., Hammond, M., Golub, D., Hirunsalee, S., Weesakul, S., Meesuk, V., Medina, N., Sanchez, A., Kumara, S., and Abbott, M. (2016). Holistic Approach to Flood Risk Assessment in Areas with Cultural Heritage: A Practical Application in Ayutthaya, Thailand. Natural Hazard, 81(1), 589-616. doi: 10.1007/s11069-015-2098-7.

Meesuk, V., Vojinović, Z., and Mynett, A.E. Extracting inundation patterns from flood watermarks with remote sensing SfM technique to enhance urban flood simulation: The case of Ayutthaya, Thailand (2017). Computer, Environment and Urban System, 64, 239-253. doi: 10.1016/j.compenvurbsys.2017.03.004.

Meesuk, V., Vojinović, Z., and Mynett, A.E. Multi-source View Surveys for Supporting Flood Management Applications (in preparation).

Conferences

Abdullah A.F., Vojinovic Z., and Meesuk V. Modelling Flood Disasters: Issues Concerning Data for 2D Numerical Models. In. International Conference on Sustainable Environment and Water Research, Meleka, Malaysia; Dec 5-6, 2016.

Paron, P, Smith M.J., Anders, N., and Meesuk, V. Quality of DEMs Derived from Kite Aerial Photogrammetry System: A Case Study of Dutch Coastal Environments. In. EGU General Assembly, Vienna, Austria; Apr 27 – May 2, 2014.

Meesuk, V., Vojinović, Z., & Mynett A. Merging Top-view LiDAR Data and Ground-view SfM Data for Enhancing 2D Urban Flood Simulations. In. 11th International Conference on Hydroinformatics, New York, United States; Aug 17-21, 2014.

Vojinović, Z., Golub, D., Weesakul, S., Keerakamolchai, W., Hirunsalee, S., Meesuk, V., Sanchez, A., Kumara, S., Manojlovic, N., and Abbott, M. Merging Quantitative and Qualitative Analyses for Flood Risk Assessment at Heritage Site, The Case of Ayutthaya, Thailand. In. 11th International Conference on Hydroinformatics – HIC 2014, New York, United States; Aug 17-21, 2014.

Meesuk, V., Vojinović, Z., and Mynett A. Merging Multidimensional Views of Remote Sensing Data for Enhancing 2D Urban Flood Simulations. In. UNESCO-IHE PhD Symposium, Delft, the Netherlands; Sep 23-24, 2013. Awarded the 1st prize for oral presentation.

Meesuk, V., Vojinović, Z., and Mynett A. Using Multidimensional Views of Photographs for Flood Modelling. In. 6th International Conference on Information and Automation for Sustainability, Beijing, China; Sep 27-29, 2012.

Meesuk, V., Vojinović, Z., and Mynett A. Using Multidimensional Views of Remote Sensing Information and Photographs for Urban Flood Modelling. In. UNESCO-IHE PhD Symposium, Delft, the Netherlands; Sep 26-30, 2011.

Netherlands Research School for the
Socio-Economic and Natural Sciences of the Environment

D I P L O M A

For specialised PhD training

The Netherlands Research School for the
Socio-Economic and Natural Sciences of the Environment
(SENSE) declares that

Vorawit Meesuk

born on 25 May 1977 in Bangkok, Thailand

has successfully fulfilled all requirements of the
Educational Programme of SENSE.

Delft, 14 June 2017

the Chairman of the SENSE board

Prof. dr. Huub Rijnaarts

the SENSE Director of Education

Dr. Ad van Dommelen

The SENSE Research School has been accredited by the Royal Netherlands Academy of Arts and Sciences (KNAW)

K O N I N K L I J K E   N E D E R L A N D S E
A K A D E M I E   V A N   W E T E N S C H A P P E N

The SENSE Research School declares that Mr Vorawit Meesuk has successfully fulfilled all requirements of the Educational PhD Programme of SENSE with a work load of 49 EC, including the following activities:

SENSE PhD Courses

o   Environmental Research in Context (2011)
o   Research in Context Activity: Co-organising two workshops on 'Disaster risk management for Phranakhon Si Ayutthaya World Heritage Site', Thailand (2014)
o   SENSE Writing Week (2013)

Other PhD and Advanced MSc Courses

o   Urban Water Systems Modelling, UNESCO-IHE Delft (2011)
o   Urban Flood Modelling and Disaster Management, UNESCO-IHE Delft (2011)
o   English for Science, UNESCO-IHE Delft (2012)
o   Advanced Academic Writing, UNESCO-IHE Delft (2012)

Management and Didactic Skills Training

o   Teaching assistant for MSc course 'GIS and remote sensing applications for the water sector' (2011-2012)
o   Teaching assistant for MSc course 'Module-Introduction to GIS' (2012)
o   Supervision of three MSc students with writing their MSc thesis (2013-2014)

Oral Presentations

o   *Using Multidimensional Views of Remote Sensing Information and Photographs for Urban Flood Modelling*. UNESCO-IHE PhD Symposium, 26-30 September 2011, Delft, The Netherlands
o   *Using Multidimensional Views of Photographs for Flood Modelling*. 6th International Conference on Information and Automation for Sustainability (ICIAfS2012), 27-29 September 2012, Beijing, China
o   *Merging multidimensional views of remote sensing data for enhancing 2D urban flood simulations*. UNESCO-IHE PhD Symposium, 23-24 September 2013, Delft, The Netherlands
o   *Merging top-view LiDAR data and ground-view SfM data for enhancing 2D urban flood simulations*. 11th International Conference on Hydroinformatics (HIC2014), 17-21 August 2014, New York, United States

SENSE Coordinator PhD Education

Dr. ing. Monique Gulickx

For Product Safety Concerns and Information please contact our EU
representative GPSR@taylorandfrancis.com Taylor & Francis Verlag GmbH,
Kaufingerstraße 24, 80331 München, Germany

Printed and bound by CPI Group (UK) Ltd, Croydon, CR0 4YY
02/05/2025
01859321-0003